Shock Waves & Man

Shock
Waves
&
Man

---

by
I. I. GLASS

Professor and Assistant Director *(Education)*
INSTITUTE FOR AEROSPACE STUDIES
UNIVERSITY OF TORONTO, CANADA

UNIVERSITY OF TORONTO INSTITUTE FOR AEROSPACE STUDIES
TORONTO 1974

Printed in Canada by
The University of Toronto Press

ISBN-09690488-0-7
L.C.74-77074

CONTENTS

To ANNE, VIVIAN, RUTH, *and* SUSAN ....

To my students, colleagues and teachers:

"Much have I learned from my teachers, even more from my colleagues, but most of all from my students".
RABBI HANINA in Ta'anith, 7a, Talmud Bavli

To the Institute for Aerospace Studies, UTIAS Quarter-Century Symposium, to be celebrated on 1 and 2 April, 1974.

ACKNOWLEDGEMENTS

It is a pleasure to thank Dr. G. N. Patterson, Founder and Director of the Institute for Aerospace Studies, for encouraging me to undertake the task of writing this book.

Research work on shock-wave phenomena at the Institute for Aerospace Studies started in 1948, and was generously supported over the years by the Canadian National Research Council, The Canadian Defence Research Board, and recently by the Ministry of Transport, Air Canada, and Atomic Energy of Canada Limited, as well as the United States Office of Naval Research (Mr. Ralph Cooper), Air Force Office of Scientific Research (Mr. Milton Rogers), Air Force Aerospace Research Laboratory (Mr. Elmer Johnson), and NASA (Mr. Ira Schwartz). Valuable additional support was received through the University of Toronto from the late Dean R. R. McLaughlin, former Dean J. M. Ham, and former Vice-President (Research) Dr. G. de B. Robinson. This generous assistance from so many quarters is gratefully acknowledged.

I was inspired to write this introductory book by stimulating and encouraging comments made by the late Dr. H. Schardin, as well as Prof. S. H. Bauer, Prof. J. H. Lee, Prof. Anatole Roshko and others, following my invited lecture on "Shock-Wave Phenomena on Earth and in Space", at the Fifth International Shock Tube Symposium, U.S.Naval Ordnance Laboratory, White Oak, Maryland, 1965. I have always appreciated their encouragement. I am grateful to Dean B. Etkin, Mr. J. J. Gottlieb, Prof. R. A. Gross, Mr. Milton Rogers, Dr. Edward Pulford, Mr. S. P. Sharma and Prof. P. P. Wegener for reading the manuscript and their helpful comments.

Finally, it is very pleasant for me to thank my colleagues at UTIAS for having selected me as the first G. N. Patterson Lecturer for the Quarter Century Symposium, 1 and 2 April, 1974. The lecture was to be based on the contents of this book.

PREFACE

Scientists and technologists are making great discoveries and proposing revolutionary theories at a rate unprecedented in human history. There is no sign that the resulting deluge of complex information is going to abate. On the contrary, the publication explosion will intensify. A greater need than ever before therefore exists for lucid reviews and popular surveys for professionals and laymen alike to interpret and assess the current state of knowledge and illuminate its impact on man and society.

This book is an attempt to provide such a survey which hopefully will give some insight into cosmic, terrestrial and man-created explosions and shock-wave phenomena. Although shock waves resulting from natural causes on Earth have invariably caused awe, wonder, horror and even tragedy, they have never posed an ultimate threat to the survival of man. With the invention of nuclear weapons in the hundred-megaton class however, all life on this planet is faced with unprecedented potential dangers from cataclysmic explosions and their associated radiation and shock waves. Even a minor nuclear exchange by warring nations could constitute a major human catastrophe.

By contrast, this volume contains many examples of significant peaceful applications of explosives and shock waves. It also outlines exciting future uses of nuclear explosives for the benefit of mankind. Their constructive utilization will undoubtedly open new avenues in providing much-needed energy resources, to replenish the supply of depletable materials and to create highly-economical industrial processes. Moreover, an understanding of cosmic explosions and shock waves promises to produce one day the key to understanding the creation of the universe, our solar system, the Earth, and even life itself.

The shock waves treated in this review arise from a very sudden release (explosion) of chemical, nuclear, electrical, radiation, or mechanical energy in a limited space. The book does not deal with cultural or sociological shocks – terms which are quite fashionable today. Societal and physical shocks do have some attributes in common. They both imply a very rapid rate of change. They both lead to a collision process resulting in a transition from an old to a new equilibrium. The duration of this transition, technically known as the relaxation time, depends on the number of disturbances that are generated. The process is irreversible. It inevitably results in a depreciation of the energy available to do useful work.

As the table of contents shows, the scope of the material covered is rather broad. It was therefore not possible or even desirable to treat each item in depth in a survey of this type. Instead, considerable reliance was placed on using many striking photographs and diagrams to complement the brief text. These illustrations have been carefully collected over the past twenty-five years. Financial considerations precluded the reproduction of some fascinating colour samples which undoubtedly would have enhanced the present contents.

Quite a number of excellent text books and numerous scientific papers already exist for the specialist on the physics, mathematics and applications of explosions and shock waves. Some of them have been listed in the references for the benefit of the professional researcher, student and lay reader. My selection of topics was influenced by over two decades of teaching, lecturing and research in the field of gasdynamics and shock-wave phenomena at the Institute for Aerospace Studies, University of Toronto, and in other parts of the world. During this period, I realized that there was a demand by non-specialists and students alike for a survey which would provide some perspective on the varied aspects of terrestrial and cosmic shock-wave phenomena and to show their relevance to our civilization. I hope that in some measure I have succeeded in accomplishing this task. If so, my labours will have been rewarded.

Toronto, 17 March, 1974 / IRVINE ISRAEL GLASS.

1

INTRODUCTION

WHAT IS A SHOCK WAVE?
WHERE DO THEY OCCUR?
HOW DO THEY AFFECT OUR LIVES?

How many times since infancy have we been startled from our sleep by the crash of thunder? How often during our waking hours have we been awed by blinding flashes of lightning, followed by the cascading crescendo of pealing thunder in a heavy downpour? The most cursory survey shows that many hundreds of lives are lost yearly throughout the world, and property damage runs into the hundreds of millions of dollars, from this type of lightning-shock-wave phenomenon alone. Still these tragedies may be dwarfed when compared with those resulting from severe earthquakes or volcanic eruptions, other examples of shock-producing phenomena on Earth. It is important to note that in the foregoing illustrations of naturally produced shock waves, the loss of life and damage to property actually result from secondary causes. In a thunderstorm, lightning channels can cause devastating fires. In a severe earthquake the shaking and heaving motions result in ruins and chaos. In a volcanic eruption the greatest havoc can come from rivers of glowing lava and flying lava bombs, burning and burying settlements remote from the explosion.

Unlike the above, the shock waves produced from chemical and nuclear explosions generate such destructive pressures, raging winds, consuming fires and deadly nuclear radiation, that life and property can be destroyed over large areas. A meteorite impact, of course, would be very similar to a man-made explosion. While there are no documented records of large numbers of fatalities resulting from impacting meteorites, one has only to contemplate the Ungava meteor crater (about two miles in diameter and about 0.3 miles deep) in the Province of Quebec, Canada, to realize that if such an impact ever struck in any of the world's large cities, its effect would be no less disastrous than the blast of a multimegaton thermonuclear bomb.

Having touched briefly upon a few of the shock-wave phenomena produced by natural events on Earth, it may be worthwhile to define what we mean by a shock wave. In descriptive terms, we can say that

FIG. 1: REPRESENTATION OF A SHOCK WAVE BY A HUGE WATER WAVE [BREAKER]
It is difficult to define shock waves in a gas, liquid or solid since their very rapid motion, even in transparent media, cannot be made visible except by using special optical methods. The huge wall of water ready to envelop the surfer is representative of the sudden pressure jump across a shock wave (Courtesy: D.H. James, Surf Photos).

it is a very sharp, thin, wave front. (It is worth noting that at sea level a shock wave in air has a thickness of about one microinch (10^{-6} inches or a millionth of an inch, whereas a wavelength of light in the infrared is about 40 microinches). A shock wave is generated when energy is suddenly released or deposited in a material (gas, liquid, solid) thereby causing an explosion. In a spherical explosion, the process is as follows: When chemical or nuclear energy is added instantly (milliseconds to microseconds) in air, water, or underground, the resulting hot, high-pressure expanding gaseous sphere drives a shock wave into the surrounding material. The shock wave races into the medium like a tidal wave and abruptly raises the

FIG. 2: THE BREAKER WAVE

A dramatic picture of a breaker wave similar to the one depicted in Fig. 1, was beautifully captured by the Japanese painter Hokusai, entitled, "View through the Waves off the Coast of Kanagawa", from his celebrated "Thirty Six Views of Mount Fuji", painted between 1823–1830.

material to a very high pressure, density and temperature. It also induces a flow velocity which races behind it. As the spherical shock wave engulfs ever-greater volumes of the material, it finally decays to a weak disturbance or sound wave (from a crash to a whisper). The decay is extremely rapid near the source of the explosion when the shock is very strong, or when the pressure, temperature, density and velocity behind it are very great, and it becomes vanishingly small as the wave becomes a sound pulse. The distance required for a shock wave to decay to a sound wave becomes progressively smaller in a gas, liquid, and solid. For example, for a given initial spherical explosive-charge diameter, the decay distance is several hundred charge diameters in air, about two in water, and a fraction of one diameter in a solid. The explosively-heated sphere expands and pulsates until it achieves an equilibrium size commensurate with the pressure of the surrounding material. The heated sphere then diffuses into the ambient air, breaks up into gas bubbles in water, or remains in the cavity underground.

Although the foregoing describes the physical nature of a shock wave, it is still a difficult event to visualize. Perhaps one of the best illustrative examples may be taken from water waves, as shown in

Fig. 1. Here we see the surfer escaping an overwhelming wall of water, which may be compared to a large abrupt pressure jump across an intense shock wave in a gas. That is, where the surfer stands the pressure is lower than at a point behind the wave at the surf-board level. Each 32-foot height of wave would add a pressure of one atmosphere at such a point. In this sense, the two phenomena are analogous. In the interior of water, or a solid, the shock wave would resemble that in a gas. Although the molecular collision processes that raise the pressure, temperature, density and flow velocity behind a moving shock wave are well understood for gases, this is not the case for liquids or solids. A significant difference stands out. Because of the large inertia of the medium, water waves break or spill at the top. Shock waves do not. It is remarkable how dramatically a similar scene was captured by the Japanese painter Hokusai in his famous painting of a breaking wave with Mount Fuji in the background as reproduced in Fig. 2.

In the case of thunder, for example, the energy is rapidly deposited in a long jagged channel by means of an electrical discharge. Since the speed of light is one million times faster than the speed of sound in air, we see the flash of a distant lightning stroke long before we hear the rumbling thunder. During an earthquake, the impulsive release of stored strain energy as the Earth's crust tears due to a shearing action, causes several types of waves to propagate within the Earth's interior and on its surface. In the case of a volcanic explosion, the rapid energy deposition occurs when high-pressure, high-temperature gases and liquids from the Earth's interior suddenly blow a large part of the volcano's peak into the atmosphere. It is estimated, for example that some cubic miles of material were shot into our atmosphere at the time of the Krakatoa volcanic explosion in 1883, a blast which caused considerable pollution and induced magnificently colourful sunsets and other upper atmospheric phenomena for many months. When a meteorite strikes the ground an immense charge of stored kinetic energy (arising from its mass and motion) is suddenly deposited at the surface. This instantly generates shock waves in the Earth as well as in the air. Similarly, when a supersonic transport produces a sonic boom in the atmosphere, the energy is provided by its engines through the fuel supply. The boom may also be felt in water or on the ground ahead of its point of impact as an earlier disturbance or precursor wave, for the sound speed in the Earth's crust or in water may exceed the flight-velocity (sound speed in water is 5000 feet per second; in soil about 3000 feet per second).

When a strong (overpressures of atmospheres) shock wave travels through an undisturbed material (gas, liquid, or solid) the pressure, temperature, and density of the disturbed state are increased many-fold. Consequently, when animals, humans or structures are hit by such a wave, they perish or are demolished as a result of the sudden violent pressure change. In addition, the aerodynamic drag (wind) produced by flow velocity can do severe damage to people, animals and structures. Even in the case of weak shock waves the pressure forces usually cause severe startle and annoyance despite the shock overpressure being as low as one thousandth of an atmosphere (two pounds per square foot). This is about the overpressure generated at ground level some distance from a lightning bolt or when a modern supersonic transport like the Concorde flies at speeds twice that of sound (1300 miles per hour or at a Mach number $M = 2$) at an altitude of fifty thousand feet. The aircraft generates conical bow and tail shock waves whose intersections on the ground resemble a horseshoe pattern which cuts a swath from fifty to one hundred miles wide across continents and oceans along its entire flight path. The resulting N-wave (so called because the pressure signature it creates over distance or time resembles the letter N) produces a disturbing "boom" or "bang". In fact, the phenomenon can be annoying enough that the question is still debated as to whether or not supersonic flights will be permitted over populated areas at all, unless the overpressures are reduced to an acceptable level by new transport designs (less than one pound per square foot). Unlike the thunderbolt, which is an instantaneous phenomenon, the sonic boom produced by the aircraft is steady in time, and each individual in its path is, in turn, subjected to a single or double boom. (N-waves shorter or longer than about 100 feet, respectively, are perceived in this way as the human ear cannot respond fast enough to differentiate both peaks of the N-wave, if it is too short).

With the invention of the bull whip, gunpowder, cannons, bombs, and fission and fusion weapons, man has learned to deposit ever-increasing quantities of energy (up to the equivalent of 100 megatons of TNT) in ever-decreasing periods of time (10^{-7} seconds or one tenth of a microsecond or one tenth of a millionth part of a second), thereby generating shock waves of almost unimaginable strength. Nuclear stockpiles have grown to the extent that there is now probably several hundreds of tons of TNT equivalent for each man, woman, and child on Earth. This unprecedented ability to overkill could turn even a minor nuclear war into a world catastrophe.

Our long and fearful association of explosives with the enormous devastation and tragedies, and the countless miseries they have wrought in wars, obscures the fact that their uses for peaceful purposes are numerous. It is doubtful if our industrial society could have been achieved without the chemical explosives so necessary for building roads, tunnels, mining, construction, metal forming and cladding, and as devices in the space program to cut, pressurize, manipulate, and accurately time the operation of various mechanisms. The future utilization of nuclear explosives for very large projects such as the building of harbours, dams, and canals, the recovery of marginal natural gas and oil deposits and geothermal heat from the Earth for multipurpose steam power plants, are only some of the many imaginative technological uses foreseen for nuclear blasts in the world of tomorrow. Controlled fission reactors (which do not involve shock waves except as the result of a possible explosion) are already generating a significant amount of electrical power throughout the world. Controlled fusion reactors, when they become a reality by the year 2000, will be able to satisfy all of the world's energy requirements for the foreseeable future.

It is also worth noting that focussed short-duration laser beams can deposit small radiation energies (a few to a few thousand joules or the equivalent fraction of one gram of TNT to a few grams) over extremely short time pulses (10^{-12} to 10^{-9} seconds; one picosecond to one nanosecond). To illustrate the miniscule value of the times in question, light travels 300,000 kilometers in one second but only 0.3 millimeters in one picosecond. Since the power is computed from the energy in joules or watt-seconds divided by the discharge time, its value can have astronomical numbers like 10^{17} watts per square centimeter. Because of this, carefully controlled and localized extremely strong shock waves and plasmas (highly conducting gas composed of a mixture of electrons, ions and neutral atoms) can be generated for study in the laboratory. Lasers may well become the "spark plugs" of controlled fusion reactors.

Even these selected examples cited show that in many parts of the world people from time to time are confronted by varied shock-wave phenomena on Earth that affect their well-being and even their lives. Similar phenomena occur in space on a truly astronomical scale, permitting us to observe these events, helping us to learn about the life and death of stars and galaxies, and to probe the evolution of the universe. Some cosmologists today believe that this evolution all started from a stupendous point-source explosion, (the "big-bang theory") or even more directly that all life on Earth owes its origins to a supernova explosion which provided the materials that made life possible.

All of this makes the study of shock-wave phenomena of great interest and importance. In many cases it will save lives to know more about this subject: for example, early-warning systems for detecting devastating hurricanes (already done through satellites) volcanic eruptions and earthquakes (still in the future), the design of structures to resist earthquakes; detection of underground nuclear explosions; prevention of mine disasters; production of heat shields to provide safe re-entry for space capsules, or for forthcoming hypersonic space shuttles and transports. Fortunately, there are many laboratory facilities such as mechanical, electrical, and chemical-driven shock tubes, hyperballistic ranges, chemical explosives for above and underground tests, and supersonic wind tunnels, where shock-wave phenomena can be studied economically in great detail and compared with analytical predictions. It is of interest to point out that the solar wind (streams of electrons and nucleons – atoms stripped of all of their electrons) from the Sun generates a natural wind in space that blows past the Earth at hypersonic speeds (Mach numbers of about ten). The Earth itself thus becomes a test model with a detached bow shock wave and tail shock wave thousands of miles thick. This flow is known as the magnetosphere flow. Scientific satellites have already measured the shock-wave structure generated in front of our Earth in this magnetosphere flow to provide important confirming data that will assist us to understand the nature of the very outermost fringes of our atmosphere. Finally, shock waves are continually being generated in the Sun and they are part of the complex processes that keep the solar nuclear furnace in operation. Without our Sun, life on Earth would not be possible, and the vital relationship between shock waves and man himself becomes evident.

2

NATURALLY GENERATED SHOCK WAVES ON EARTH

Some of the natural shock-wave phenomena which one can encounter on Earth were touched on briefly in the Introduction. Now they will be examined in greater detail. It is worthwhile to reconsider, first, the type of explosion that is generated by lightning by looking at a small cylindrical portion of the electrical discharge. Roughly similar results are obtained, for example, from an explosion of a sphere of TNT. This is illustrated in Fig. 3 for three consecutive times. It is seen under varying conditions of pressure p with radius r, from the shock wave to the boundary (contact surface) of the driving, expanding gas (whether heated electrically by an electic discharge, chemically by an explosive, or radiantly by a laser beam), that the shock wave pressure decays as distance from the explosion increases. The decay does not take place at a constant rate. It is greatest at early times and least at late times (a nonlinear process). The pressure profile in the interior of the driving gas is more complex, and for the sake of simplicity, is not shown. In practice, the contact surface motion is also complex, achieving an equilibrium position and then dissipating through convection and diffusion. This simplified picture of lightning-generated shock waves can be experimentally verified by observing the discharge through an exploding wire, as shown in Fig. 4. The vaporization of the wire due to the passage of a high-voltage current can be seen to form a glowing ionized channel. The expanding ionized-copper vapour drives a blast or shock wave before it that decays to a sonic boom sounding very much like thunder. In fact the channel can be regarded as a slice from a larger lightning discharge channel occurring during a storm (see Fig. 7 for further details).

Our eyes cannot normally distinguish a shock wave in a gas, liquid, or solid. For example, air is invisible to us whether it is at low or high pressure (density). However, as the density of a gas increases, the refractive index, an optical property, also increases. This makes it possible to visualize shock waves by means of relatively simple techniques such as interferometry, schlieren, and shadow photography. An interferogram shows the density change across a shock wave; a schlieren photograph yields a record of the rate of change of the density, that is, its gradient; and a shadowgram records the rate of

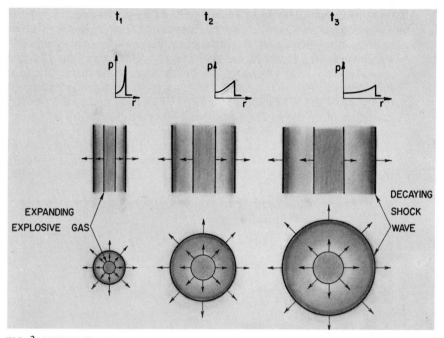

FIG. 3: SCHEMATIC DIAGRAMS OF CYLINDRICAL AND SPHERICAL EXPLOSIONS

The exploding gas drives a steep-fronted blast wave whose pressure (p) decays with distance (r). The explosive gas achieves an oscillatory finite size, whereas the volume of air engulfed and heated by the shock wave grows with distance, and in this manner the shock wave spends its energy and decays. The shock wave decay rate near the source of the explosion is most rapid for spherical, cylindrical and planar explosions, respectively. Far away from the source the shock waves all decay to sound waves. The diagrams illustrate the explosion at an early time t = t₁ and at later times t₂ and t₃. (Courtesy: UTIAS).

change of the gradient itself. Characteristically, the shock wave appears as a black line followed by a white line in a shadowgram as shown in Fig. 4 (and much more clearly in Fig. 5).

Figure 5 is a shadowgram of a bursting sphere initially filled with air at high pressure (300 pounds per square inch, psi), which is ruptured by a plunger and photographed 375 microseconds after impact, with an exposure time of only a few microseconds. Due to initial high pressure, the gas rushes out through the spreading cracks in the sphere. This escaping gas, which has now been disturbed in its outrush by the glass fragments, is turbulent and asymmetric, yet it drives ahead of itself a beautifully symmetric shock wave. The black and white structure of the shock wave noted above (as well as of the turbulent driver gas) is clearly shown.

The same process is illustrated in Fig. 6, using an electric spark for a light source to yield the schlieren photograph of a blast from a bursting 2-inch-diameter glass sphere initially pressurized at 550 psi. The photograph was taken 100 microseconds after the sphere started to break up. Once again, the turbulent driver gas, and the symmetric shock wave in front of it, are seen clearly. This time, however, the gradations on the left-hand side of the photograph are just opposite to those on the right-hand side. This is typical of a schlieren record produced when a vertical knife edge is used to reveal the density gradients of the flow. Unlike a shock-wave shadowgram, which appears as a black line followed by a white line, a schlieren record shows the shock as either black or white, depending on the direction of motion with respect to the knife edge. (Further details can be found in the book by Liepmann and Roshko listed in the References). Explod-

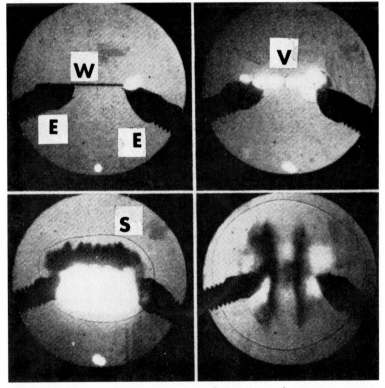

FIG. 4: PHOTOGRAPHS OF A BLAST WAVE (SHOCK WAVE) GENERATED BY AN EXPLODING COPPER WIRE

The copper wire was 0.002 inches diameter. The electrical condenser used in providing the explosive energy was charged to 2000 volts. The exposure time for each picture was 10 nanoseconds (1/100 microseconds). The framing sequence from upper left to lower right is 5.4, 5.7, 10.0 and 15 microseconds. E = electrode, W = wire, V = vaporized wire, S = shock wave. (Courtesy: Beckman and Whitley).

ing a sphere of TNT or other chemical explosive would produce a process very similar to that shown in Fig. 6, that is, the explosive is converted into a high-pressure high-temperature gas driving a shock wave in front of it. The gas would eventually attain an equilibrium volume and the shock wave, no longer being driven by an expanding gas, would decay with propagation distance until it became a sound wave. (See, for example, Fig. 22).

Thunder

It has already been said that thunder is probably the most prevalent example of a shock wave that man encounters during his span on the

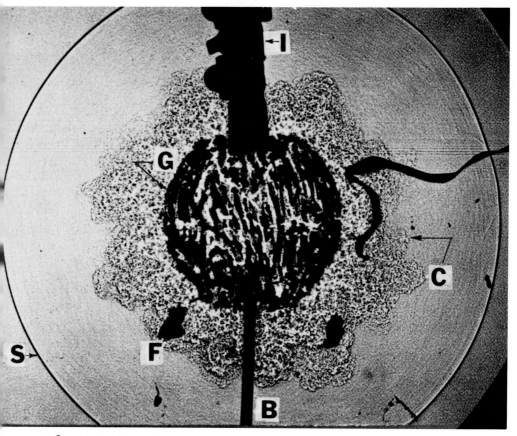

FIG. 5: SPARK-SHADOWGRAM OF AN EXPLOSION

The explosion was generated in atmospheric air by bursting a 2.0 inch diameter glass sphere pressurized with air to 20 atmospheres (300 psi), and photographed 375 microseconds after its rupture by a mechanical breaker (B). I = pressure inlet pipe, G = broken glass sphere, F = glass fragment, C = turbulent contact front, S = shock wave (Courtesy: UTIAS).

Earth. From infancy to old age the claps, peals, crashes, rolls and rumbles of thunder never fail to produce a startling and awesome effect, especially close to the discharge point, where one of the loudest of all natural sounds is generated. A dramatic and outstanding

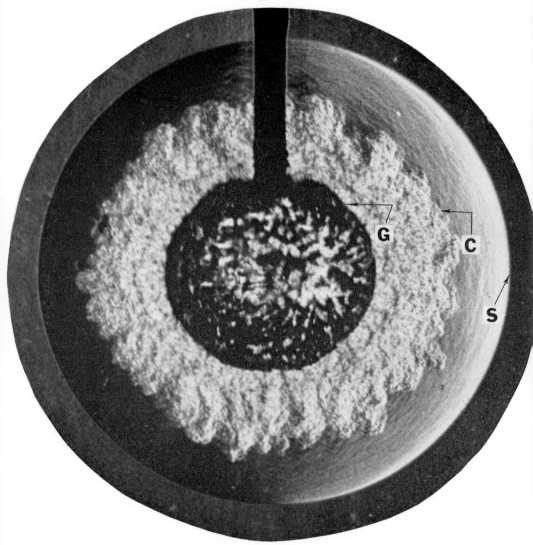

FIG. 6: SCHLIEREN PHOTOGRAPH OF AN EXPLOSION

The explosion was generated in atmospheric air by bursting a 2.0-inch-diameter glass sphere pressurized with air to 37 atmospheres. The photograph, with a 2-microsecond exposure, was taken 150 microseconds after rupture through over-pressure. G = broken glass sphere, C = turbulent contact front, S = shock wave. (Courtesy: UTIAS).

photograph of lightning discharges appears in Fig. 7. Several brilliant, electrically heated channels and streamers stand out. Owing to the 45-second time exposure, the recent ones are thinner than the later strokes (see Figs. 3 and 4). The initial channel radius is estimated to be only a few millimeters. It grows to a few centimeters in about 30 microseconds as the instantaneously added electrical energy heats, pressurizes and ionizes the air in the exploding channel. The temperature in the channel remains at about 20,000°C, and the pressure drops from about 35 to 5 atmospheres in this period. Peak pressure values are even greater. It is obvious why close proximity to the channel can be hazardous.

The meandering paths of the channels and streamers as the discharge oscillates from the clouds to the ground can also be seen. Paths such as these contribute to the rolling and rumbling sensations of thunder. Fortunately, little damage was caused. However, with channel temperatures of many thousands of degrees, and pressures of several atmospheres, severe damage to buildings and loss of life can occur. Some channels may be several miles long and could have currents of hundreds of thousands of amperes. Because of the extreme temperatures, the air molecules are split apart (dissociation) to form nitrogen and oxygen atoms, and some of the atoms have electrons torn from their shells (ionization). It is this mixture of electrons, ions and atoms that is known as a plasma, and the gas, which originally was a poor conductor now becomes a very good electrical conductor.

Benjamin Franklin correctly established the nature of lightning as an electrical spark in 1752 and invented the lightning rod for the protection of buildings. However, although the penomenon of thunder has occupied the minds of men throughout the ages, much remains to be learned about its electrical, thermal, and acoustic properties.

It is estimated that over sixteen million thunderstorms occur on Earth each year, and that about 100 lightning discharges are taking place somewhere in the atmosphere every second. Some parts of the world have many thunderstorms and others very few (222 days of the year for Java and about 4 per year for the California coast). Severe storms, tornadoes, and hurricanes cause many deaths and terrible destruction. For example, the great cyclone and huge tidal waves that laid waste to parts of East Pakistan (now Bangladesh), in 1970 claimed the lives of from 500,000 to two million people, and may well be one of history's worst human disasters. Damage in the u.s.a. from such causes approaches one billion dollars per year. Fortunately, satellite monitoring techniques have saved tens of thousands of lives by giving sufficient advance warning to permit evacuation of populated areas of those coastal regions that lie in the paths of hurricanes. Such early-warning systems and methods for diminishing the strength and changing the path of great thunderstorms will bring relief from such disasters.

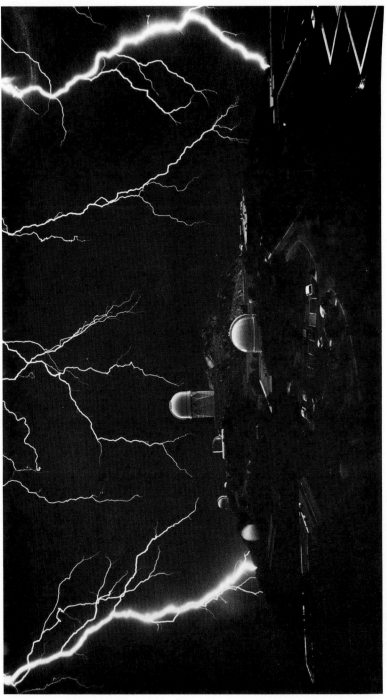

FIG. 7: SHOCK WAVES FROM LIGHTNING
A dramatic presentation of lightning strokes over Kitt Peak National Observatory, Arizona. The 45-second exposure shows lightning channels just generated (thin) and those in expanding motion (thick). The channels drive blast waves or shock waves, which can be very damaging near their hot, high-pressure cores. When one is near them, the shock wave has a frightening crack; far away the jagged channels produce the rumble of thunder. (Courtesy: © Gary Ladd 1972).

Earthquakes

A schematic diagram of what is now believed to be the structure of the Earth's interior is shown in Fig. 8. The core (2200 mile radius) is believed to be of nickel-iron. It is liquid outside and solid inside at temperatures of perhaps 7500°F. The mantle, which is usually considered to be a solid since it can transmit shear waves, occupies the second largest portion of the globe. It may have a temperature of 7000°F. Since liquids cannot transmit shear waves, it has become possible to determine the internal structure of the Earth (and recently the Moon). The relatively thin crust of about 5 to 40 miles (average about 10) forms a hard outer shell. Owing to the pressure and thermal-induced forces, the entire system is in continual motion,

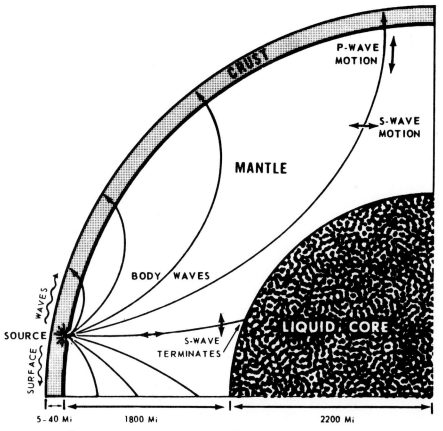

FIG. 8: THE WAVES GENERATED BY AN EARTHQUAKE OR EXPLOSION

During an earthquake or underground nuclear explosion, the energy released by the source produces several types of waves: surface waves, S (shear waves) and P (push-pull or longitudinal) waves. A knowledge of these waves makes possible the prediction of the position and strength of the source. (Courtesy: Batelle Institute).

albeit slight in some parts. It was once thought that all earthquakes
originated in the crust, as shown in the diagram. However, opinion
now appears to favour the mantle as the source of most earthquakes.

The stresses produced in the mantle by existing great pressures or
from underground nuclear explosions form the source or focus of the
earthquake, which causes a shaking of rock or earth tremors. Accel-
erations of 1 g have produced shearing motions in surface levels of
from 1 to 35 feet. The damage and loss of life can be truly catastrophic, as Fig. 9 illustrates graphically. The loss of life is usually due
to the shattering of buildings and houses into rubble, killing the people
indoors and in the streets. It is doubtful if any loss of life has resulted
directly from real shock waves in the ground, or those that may be
induced in the air. It is estimated that 800,000 people were killed on 24
January, 1556 in Shensi, China; 300,000 in Calcutta, India, on 11
October, 1737; and 60,000 in Lisbon, Portugal, on 1 November, 1755.
Many more perished in the rest of Portugal and Spain in the latter
event, which was probably the most violent earthquake in historic

FIG. 9: DEVASTATION PRODUCED BY EARTHQUAKES
The ruins left by an earthquake, that shook several towns of western Sicily in January
1968, are searched for survivors. The devastation is a grim reminder of the awesome
power of these natural phenomena. (Courtesy: Paris Match).

times. About 85,000 died in Messina, Italy, on 28 December, 1908; 95,000 in Tokyo-Yokohama, Japan, on 1 September, 1923, and more than 12,000 in Iran on 1 September, 1962. These are only a few of the most devastating earthquakes on record. It is estimated, however, that they have taken a total toll of 14 *million* people throughout recorded history. Fortunately, out of about 100,000 shocks that occur yearly, only about 10 will produce major earthquakes. Even so, one great earthquake in a given year can release more energy than all the rest combined. The magnitude of the ground motion in an earthquake was first satisfactorily defined by Prof. C.F. Richter. For example, a magnitude 7 earthquake on the Richter scale represents a major earthquake with an equivalent energy release of about one megaton of TNT (or a force that may be likened to 50 of the bombs used on Hiroshima). Only about 25 such earthquakes occur per year. The greatest recorded magnitude is about 9 on the Richter scale; the equivalent of about one thousand megatons, or 10,000,000 railroad cars of TNT! Underground explosions in the megaton range have a Richter-scale range of 7 and up, and in the kiloton range of 5 and up.

In a nuclear explosion (see Chapter 5), the high-pressure (millions of atmospheres), high-temperature (millions of degrees centigrade), fireball produces an intense shock wave in the Earth capable of vaporizing the rock around it. The shock wave then decays quickly with distance. However, it is not certain whether shock waves are also produced at the source of an earthquake. Although a very great release of energy is involved in a large earthquake, unlike an underground nuclear explosion, it is spread over large volumes and strong shock waves may not be produced during every earthquake. However, in both cases, the initial types of waves are transformed into classic elastic (P, longitudinal; and S, shear) and surface waves. In a nuclear explosion the initial motion recorded on a seismograph is always away from the source; whereas in an earthquake it can also be towards the source. Wavelengths and amplitudes from both sources also differ in magnitude (explosions have lower surface wave amplitudes). However, there is no absolutely certain way of differentiating between an explosion and an earthquake if the source is located in an earthquake region such as, say, the Aleutian Island chain. Aside from this remote possibility, seismologists today feel confident that an underground nuclear-test-ban treaty could be supervised if once negotiated, and would provide a valuable step towards eventual total disarmament.

P-waves are compression and expansion disturbances in the direction of the wave path (push-pull or longitudinal waves). S-waves displace the particles at right angles to their paths and are known as transverse, shear waves, or shake waves. Both of these waves, also called body waves, go through the Earth's mantle and crust. From a study of both types of paths, seismologists can predict the focus, hypocentre or source of the earthquake or thermonuclear blast and the point directly above the focus on the surface of the Earth, the epicentre. The surface waves are analogous to those generated in water, and two or three types have been identified. They have a speed of about two to three thousand feet per second, depending on the type of soil. However, the push-pull waves in the Earth's interior have the highest velocity (6.3 miles per second) and arrive at a point first. They are followed by the "shake" or secondary waves (at 3.5 miles per second). The fact that the S-wave is not reflected from the core indicates that the core itself must be liquid or gas and provides a measure of its depth, as shown in Fig. 8. However, the mantle is "rigid" enough to transmit the P and S-waves. It should be noted that the motion produced by the waves decays with distance from the source (earthquake or underground explosion) as its energy is dissipated over an ever-increasing mass of material.

Many sounds are associated with an earthquake, depending on one's proximity to the source. Near the source, the impressions are sharp like the tearing of blocks of rocks; farther away like a clap of thunder, or boom of an explosion; sometimes, no sound is heard. Landslides and avalanches can take place in rocky terrain. When these occur underwater, great disturbance waves (tsunamis) are generated. When the tsunamis hit a rocky shelf along a shoreline, they are slowed down. However, the huge mass of water above the shelf continues to rush forward, thereby generating great "tidal waves". Tsunamis of over 200 feet in height which have caused great damage in their powerful lunge onto the shore have been recorded. A tsunami in the open sea has a height of less than a foot, but a length between crests that may approach one hundred miles. It is the enormous mass of this water, sometimes moving at velocities over 600 feet per second, that provides the wave's gigantic destructive energy. Surface waves, in the form of "storm tides" associated with hurricanes and typhoons in areas of sharp barometric lows, have a similar appearance and can be equally devastating (as was the Bangladesh [East Pakistan] disaster of 1970). Neither a tsunami nor a storm tide is a tidal wave, since they are not associated with the Earth's tides. The latter, of course, are responses to Lunar (and Solar) gravitational forces.

Some parts of the world are more earthquake-prone than others. For example, steps are already being taken in Tokyo to reduce the casualties from a possible disaster that is expected to occur in the 1990's, an upheaval similar to that which occurred in 1923. When one considers that there are about 12 million people in the Tokyo-Yokohama area, such an earthquake could cause a million casualties. It is difficult to comprehend the magnitude of such a catastrophe.

It is also worth noting that a recent study of the State of California indicates that, over the next three decades, the region may sustain as many as 50,000 deaths, fourfold as many hospitalized injuries, and property damage of 20 billion dollars due to earthquakes. Obviously, research in this field is of vital importance if these staggering potential losses are to be minimized. Recent studies show that predictions and early warnings based on scientific premonitory signals are a distinct possiblity in the near future.

Volcanic Eruptions

The ejected steam, burning gases and solid material in volcanic eruptions can be spectacular, and the shock waves generated by this fast-moving mass can be heard over great distances. The eruption at Krakatoa, an uninhabitated volcanic island between Java and Sumatra, is a good example. Figure 10 shows the early stages of the eruption a month before the tremendous explosion in May of 1883, and a map of the audibility of the shock waves during 26–27 August, 1883. The circles from the source are about 700 miles apart and indicate that the blast was heard over distances exceeding 3000 miles. It is estimated that 1-1/8 cubic miles of material was hurled into the atmosphere, requiring an energy equivalent of 5000 megatons of TNT (250,000 bombs of the type used on Hiroshima). The explosion re-

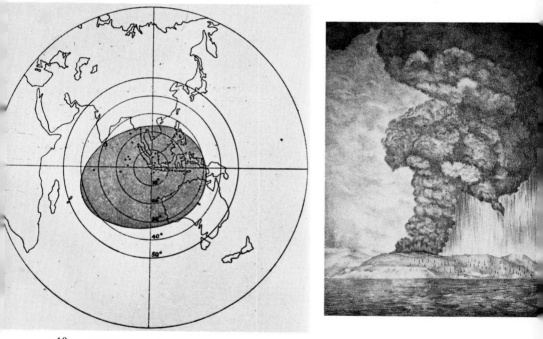

FIG. 10: THE KRAKATOA EXPLOSION

A view of Krakatoa during the earlier stages of its eruption as shown in the sketch (right sketch) was made on Sunday, 27 May, 1883 (note the resemblance to the photograph of Fig. 11). The actual explosions took place on 26–27 August, 1883, with an energy release equivalent to 5000 megatons of TNT or 250,000 bombs of the type used on Hiroshima. It was heard 3000 miles away (shaded area, left sketch). A mass of over a cubic mile of debris was shot out and a tidal wave 150 feet high caused the death of 36,380 people in neighbouring Java. (Courtesy: Royal Society of London).

sulted in the total demolition and disappearance of two-thirds of the island. The volcano, which was originally 2700 feet high, disappeared below sea level, and consequently, an immense tsunami 150 feet in

FIG. 11: SPECTACULAR VOLCANIC EXPLOSION IN ICELAND
The volcanic inferno and boiling sea generate a 30,000 foot billowing column of smoke and steam over Iceland's Heimaey Island. All of its 5000 inhabitants were safely evacuated although the damage was great. (Courtesy: National Geographic Society).

height was generated and moved on to western Java, where it killed over 36,000 people.

When the volcano blew its top, minute dust particles were shot to altitudes of over 130,000 feet. The dust changed the radiation pattern received from the Sun, thus lowering the Earth's temperature, as well as causing unusually brilliant sunsets all over the world.

A more recent example of a violent volcanic eruption is shown in Fig. 11. It shows a billowing column of smoke and steam rising (sometimes to 30,000 feet) from the glowing volcano and the boiling lava as it falls into the sea off the small island of Heimaey, six miles from the south-west coast of Iceland. It forced the evacuation, on January 1, 1973, of 5,000 inhabitants from the town of Vestmannaey-jar, Iceland's most important fishing village, as the "lava bombs" burned and buried it. Although tragic dislocations and misery re-sulted for many people, not a single life was lost in the evacuation of these stalwart villagers.

The driving energy for volcanic eruptions comes from the bottled up steam and gases generated by the molten rock (lava) produced by hot, high-pressure material within the Earth's mantle. Although shock waves from such a source may sound fearful and awesome, very few recorded casualties have resulted from its overpressures or induced winds. It is the hot lava (1300°C) that causes spectacular fires and leaves death and destruction in its wake.

It has been estimated that about 430 volcanoes in the world have erupted at least once within historic times. Of the 2500 known erup-tions, 2000 took place in the Pacific Ocean area, where 336 active volcanoes still exist.

Meteorite Impact

It is possible to encounter meteoroids (space debris in orbit) travelling with speeds varying from the Earth's escape velocity (7 miles per second) to speeds of escape from our solar system (45 miles per second). If it is true that the meteoroids come from the asteroid belt between Mars and Jupiter, then velocities of intermediate values in this range can be expected. However, as the meteoroids pass through the denser layers of the atmosphere, they are decelerated by drag and heated to incandescence by shock-wave compression and friction. The smaller particles burn up and give rise to the well-known shoot-ing stars and the larger masses are heated to brilliant and spectacular glowing bodies followed by an evaporated (ablated) luminous and smokey tail as seen against the sky. (Figs. 12 and 13).

The velocity at impact with the ground (when a meteoroid impacts it is called a meteorite, the word meteor is also used in the literature for both terms) often appears to be only about 700 feet per second, less than sound velocity in air (1130 feet per second). Otherwise a ballistic supersonic shock-wave pattern would be formed in the denser layers of the atmosphere that would give rise to a sonic boom akin to that caused by a bullet or shell. It is worth noting that a sonic boom would still be heard, as is the case for a space-capsule re-entry, for a larger meteorite, even if its speed was subsonic at impact. The shock wave generated when it was travelling at supersonic speeds at higher altitudes would still propagate towards the ground and be heard as a boom. It is now believed that the Tunguska meteoroid which exploded over Siberia in 1908, was made up of cometary ices and that it disintegrated before striking the ground. Nevertheless, the shock wave which was initially generated on entry into the Earth's atmosphere was so strong that 800 square miles of forest was burned and flattened. It was very fortunate that in this remote area no lives were lost due to the fury of a shock wave whose overpressures and wind dynamic pressures were probably in the range of hundreds of pounds per square foot (fractions of an atmosphere). In this regard the meteoroid entry was similar to an intense above-ground explosion. It is estimated that the energy release was equivalent to between 25 and 250 megaton bombs.

Owing to their high kinetic energies, the velocities could be very great for large meteorites. Ballistic shocks (as from a shell) would then occur on entry. New shock waves would be generated on impact in the ground and in the air as the kinetic energy of the meteorite is suddenly released explosively. In this process a good part of the meteorite, and the ground where it strikes, are evaporated and much debris is thrown out to form a crater. Figure 14 shows the Arizona

FIG. 12: OBSERVATION OF A METEORITE IMPACT
An artist's conception, from eyewitness accounts, of a meteorite impact at Agram or Hraschina, Yugoslavia in 1751. Two iron meteorites, 88 and 20 pounds respectively, were recovered. (Courtesy: *Selected Topics on Ballistics*, AGARDOgraph 32 edited by W. C. Nelson, Pergamon Press, New York, 1959).

crater estimated to have been produced by a meteorite mass between 10 thousand and 100 thousand tons between 5 thousand and 50 thousand years ago. If it is assumed that the incoming meteorite weighed 60,000 tons (this would be equivalent to a nickel-iron sphere 80 feet in diameter with a specific gravity of 8), and that the entry velocity was equal to the escape speed, then the TNT equivalent would have been equal to one megaton bomb. The crater has a diameter of about 4300 feet (0.8 miles) and a maximum depth of about 560 feet (0.1) miles). It is frighteningly obvious that if such an impact occurred in a major world city such as New York (see Fig. 14), Tokyo, or London, the effects would be as disastrous as those from a

FIG. 13: SPECTACULAR METEOROID TRAIL OF SMOKE AND FIRE
The fall of the iron meteorite of Sikhote-Alin, north of Vladivostok, U.S.S.R., February 12, 1947, about 10:38 a.m. produced this dramatic luminous and billowing path. (Courtesy: *Meteorites* by Fritz Heide, University of Chicago Press, 1964).

megaton nuclear bomb. In this case, destruction would occur not only from the upheavals of the impact but also from the intense shock wave, with overpressures of many atmospheres, and very large wind velocities, which would crush and demolish everything in their path over a large area.

The largest, readily-recognizable crater on Earth is the Ungava (Chubb or Meen) crater, at the northern tip of Quebec, Canada (Fig. 15). It has a diameter of about 11,500 feet (2.2 miles) and a maximum depth of 1430 feet (0.27 miles). It is estimated that at impact the giant meteorite had a hypervelocity greater than 8.7 miles per second, and after it struck it "left the surrounding barren plain a chaos of boul-

FIG. 14: THE CATASTROPHIC DAMAGE POTENTIAL FROM METEORITES

The top photograph shows an outline of the Arizona Crater (Canyon Diablo), 4300 feet in diameter and 560 feet deep, which appears in an aerial photograph below, superimposed on the southern tip of Manhattan Island (New York). An event of this type in a modern city would be as disastrous as a thermonuclear explosion. (Courtesy: *Meteorites* by Fritz Heide, University of Chicago Press, 1964).

FIG. 15: THE UNGAVA (CHUBB OR MEEN) CRATER IN NORTHERN QUEBEC, CANADA
The crater is about 2.2 miles in diameter, 0.27 miles deep, and is now filled with water. The meteorite fractured and sprayed an estimated 5 billion tons of shattered granite over the area. (Courtesy: © National Geographic Society).

ders''. The ejected debris has been estimated as five billion tons of shattered granite.

Present studies indicate that there may be as many as 50 circular features on Earth, ranging up to 37 miles in diameter, that were caused by meteorite hypervelocity impact. Lunar craters are many times larger and deeper and orders greater in number. The origin of the craters on the Earth, the Moon and Mars (Figs. 16 to 19) is now believed to be due to meteorite impact. However, this is still the subject of much speculation, despite the manned exploration of the Moon and the unmanned exploration of Venus and Mars. The latter has two badly battered and cratered moons, Deimos and Phobos, which are extreme examples of cratering.

It is worth mentioning a recent meteorite impact on the Moon that took place on 17 July, 1972. From the Apollo seismic stations left on the Moon and transmitting to Earth, it was estimated that the meteorite weighed about a ton. It impacted at 5 miles per second. The kinetic energy release on impact was equivalent to 7 tons of TNT, or 28 tons of TNT if the speed was 10 miles per second.

Great interest can also be expressed in an impact that took place about 4 billion years ago. It is believed that a meteorite of 20 or even 50 miles in diameter, travelling at between 10 and 20 miles per second, hit the Moon. The shock waves and rarefaction waves generated in the meteorite caused it to disintegrate and vapourize. The shock waves and rarefaction waves in the Moon caused a sea of debris, molten rock and a ring of mountains 650 miles in diameter. The bowl crater was filled with lava percolating through from the interior to the mantle and crust. It is known today as the Imbrium basin, the large "eye" of the man in the Moon. The energy release is estimated as the equivalent of billions of hydrogen bombs, the flash of light generated by the impact momentarily rivalled the sun in intensity. The ejecta scattered a geological blanket for 1000 miles in every dirction, thicker than one mile near the crater and some feet at the outer edges. Some ejecta escaped the Moon into outer space, and as some scientists believe, perhaps even hit the Earth in the form of rocky fragments or tektites.

The Moon like the Earth has an established age of 5 billion years. Its origin is still debated and remains a mystery. It has no active volcanoes and apparently no valuable resources worthy of exploitation.

FIG. 16: THE BATTERED AND CRATERED FACE OF THE MOON
This beautiful and dramatic photograph of the gouged and pock-marked lunar surface was taken by the crew of Apollo 17, astronauts Cernan, Schmitt and Evans, on 24 December 1972 during the last of the lunar missions. The sight of the "new" blue Earth in the lunar skies was always a source of comfort to them. (Courtesy: NASA).

FIG. 17: A CLOSE-UP OF A LUNAR CRATER
Alan Shepard, Commander of Apollo 14, with his Modular Equipment Transporter stands near a lunar crater. The lunar landings have added greatly to our knowledge in this field, although a definitive theory for the origin of lunar craters is yet to be evolved. (Courtesy: NASA).

FIG. 18: THE CRATERS ON MARS

A photograph from Mariner 6 (N22) taken during July, 1969, surprisingly shows the cratered surface of Mars illuminated by a low Sun (14-degree angle above the horizon). Unlike the Moon, there are two distinct classes of craters: small, young, bowl-shaped craters and flat-bottomed craters (modified by unknown processes) as shown clearly at the bottom of the photograph. (Courtesy: NASA).

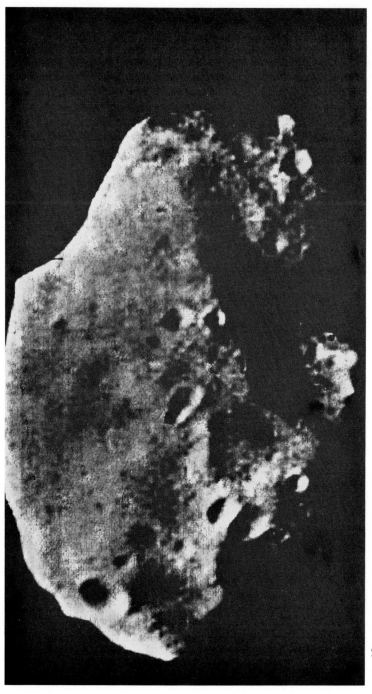

FIG. 19: AN UNROMANTIC-LOOKING MOON
Phobos, the inner moon of Mars, is a chunk of rock about 17 miles long by 25 miles wide which has been badly battered and cratered. Deimos, the other satellite is about half the size of Phobos. The photograph was taken by Mariner 9 from a distance of about 3500 miles. (Courtesy: NASA).

3

ARTIFICIALLY GENERATED SHOCK WAVES ON EARTH

In the previous section, some examples were given showing how shock waves generated from natural causes affect people's way of living and take their toll of life in many different geographical localities. Then, man progressively learned how to generate and increase the strength of shock waves by inventing the bull whip, gun powder, improved chemical explosives and, finally, nuclear and thermonuclear bombs. With these latter devices he could rival or even surpass the destruction caused by some of the natural-disaster phenomena on Earth. So much so, that today, our very existence is threatened by the size of the "overkill" of the nuclear arsenals of the world powers. Unlike some of the cataclysms of nature, death and destruction due to man-made blasts come directly from shock-wave overpressure and nuclear radiation.

Bull Whip

Perhaps one of the earliest methods of generating weak shock waves was by using a whip. The sharp, startling crack was quite common years ago in our agricultural society. Now we must attend the circus to hear the unique cracks from the animal trainers' whips. Similar cracks or bangs can also be produced quite readily by bursting a pressurized balloon or a paper bag or by firing a toy compressed-air cork gun. All of these devices produce their own characteristic booms which can be measured quantitatively by using recording microphones or pressure gauges.

Figures 20 and 21 show how a shock wave was produced by cracking a 12-foot bull whip. The operation requires considerable skill in transferring a man's muscular energy into a travelling loop (see Fig. 20). In negotiating the travelling loop, the tip-threads are momentarily accelerated to about 1400 feet per second and act as a driving piston to produce a shock wave. In the photograph, the wave momentarily travels at a speed of 2200 feet per second (twice the speed of sound) or at a Mach number $M = 2$, and quickly decays in strength. This process is well illustrated in Fig. 21. The shadowgram shows how the accelerating tip actually produces the shock wave in frames 2 to 4.

Gunpowder

With the invention of gunpowder and modern chemical explosives, man was able to increase his capacity for generating ever stronger shock waves for destructive or peaceful purposes. It appears that gunpowder or blasting powder was first used in the Thirteenth century. Its use in coal mines in Saxony was recorded as early as 1627,

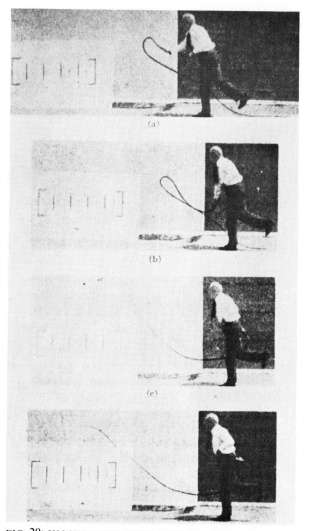

(a)

(b)

(c)

FIG. 20: SHOCK WAVES GENERATED BY A BULL WHIP

The crack heard from a whip is a shock wave. The details of their production by a 12-foot bull whip are shown in the above four photographs at 120 frames per second (and even more precisely in Fig. 21). (Courtesy: Naval Research Laboratory).

and in Cornwall in 1689. It came into general use in 1831, after the invention of the safety fuse.

Alfred Nobel made several notable contributions between 1867 and 1879, which included development of the fulminate detonator, dynamite, blasting gelatine and explosives of varying strengths. These inventions were followed by new ones from other innovators, which included more sophisticated types of explosives, electric detonators, detonating fuses and shaped charges. In the military field, explosives and propellants with slow burning rates, are used as initiators and accelerators in all types of ammunition for rifles, guns, rocket launchers, grenades, bombs, torpedoes, ballistic missiles, and demolition work.

In industrial applications for peaceful purposes, explosives are used extensively in mining, civil engineering, agriculture, mechanical engineering, and in space programs. Typical uses in mining are excavations for shafts, tunnels, headings, and galleries; recovery of minerals from mines, quarries and collieries, and geophysical explorations. Some good examples in civil engineering are tunnel excavations for railways and roads, and for digging building foundations,

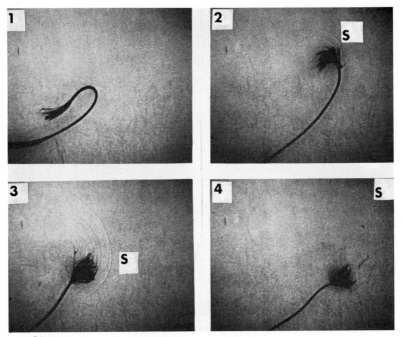

FIG. 21: SHOCK WAVES FROM THE FLICK OF A WHIP

The motion and shape of the tip of a bull whip and the production of shock waves, (tip speed 1400 feet per second; shock speed 2200 feet per second; sound speed 1100 feet per second) are clearly shown in the four shadowgram frames. (Courtesy: Naval Research Laboratory).

canals, channels and harbours. Agricultural applications for explosives can be found in well-sinking, irrigation and drainage ditches, removal of tree stumps, fence-post sinking, boulder blasting and loosening of subsoil for orchards. In the space program, explosives (pyrotechnics) are used to time, actuate, pressurize and cut various devices at certain times during the entire duration of the mission from takeoff to recovery. In a complex system such as the Saturn-Apollo combination, there are dozens of functions performed with great precision by explosives during the normal course of operation or in an emergency when the usual type of mechanical system may become inoperative. The Apollo 13 mission was saved by just such an explosive device in freeing the control module from the lunar excursion module.

FIG. 22: EXPLOSION FROM A 20-TON HEMISPHERE OF TNT
The blast wave S, and fireball F, from a 20-ton TNT surface explosion are clearly shown. The backdrops are 50 feet by 30 feet and in conjunction with the rocket smoke trails, it is possible to distinguish shock waves and particle paths and to measure their velocities. Owing to unusual daylight conditions, the hemispherical shock wave became visible. (Courtesy: Defence Research Board of Canada).

It is worthwhile quoting from Wardell (1892) in this regard:

"It lies beyond the subject of this article to attempt an estimate of the influence, direct or indirect, upon modern civilization of the introduction of explosive agents for the purposes of war. Some eminent authors have gone so far as to consider the invention of gunpowder as next in importance, in the ultimate effects, to those of printing and the application of steam power. However, this may be, it is well to remember that explosive substances are now of immense utility in the arts of peace: indeed it is not too much to say that without their aid many of the great engineering enterprises of the present day would either be impossible or else have to be carried out at a vast additional expenditure of time and labour."

Some representative photographs, which illustrate a few of the important properties of chemical explosions, are shown in Figs. 22 to 24. Figure 22 shows the detonation of a 20-ton TNT hemispherical charge at the Defence Research Board's Suffield test station at Ralston, Alberta, Canada. The turbulent white explosive gas can be seen driving a hemispherical shock wave in front of it, very much like the bursts shown in Figs. 5 and 6. The white streamers are rocket smoke trails used to visualize the position of the shock wave and the particle paths which are induced behind it. The shock wave appears as a shadow-like effect produced by the natural light existing at the time of explosion. Figure 23 shows the actual construction of a 500-ton charge of TNT, and an aerial photograph of its subsequent detonation. Figure 24 depicts the fireball at a later stage and the position of the reflected blast wave as indicated by the break in the streamers on the right. Even larger charges of 1000 tons have been detonated. In all of these cases, physical measurements were used to verify analytical predictions for very large (nuclear) explosions, which could not be tested in practice owing to the partial nuclear test ban treaty.

The shadowgrams shown in Figs. 25 to 28 illustrate the formation of shock waves which produce the sudden crack or "sonic boom", from a 0.30-inch-diameter bore rifle at 75, 175, 250 and 400 microseconds after a bullet has been fired. Figure 25 shows how the plane shock wave generated by the moving bullet, which compresses and heats the air in front of it in the barrel, comes out of the muzzle and diffracts to become a spherical shock wave. The shock-compressed and heated gas comes out of the muzzle as a supersonic jet. In Fig. 26 the bullet is just leaving the muzzle, followed by the hot high-pressure gas generated by the explosive charge. The jet outflow has become more turbulent with time and thereby causes the generation of weak shock waves throughout the flow field. A secondary shock that has caught up with the first emergent wave may be seen. Figure 27 shows the beginnings of the strong blast wave driven by the explosive gas. It is moving rapidly outward to overtake the primary shock wave and, in doing so, envelops the bullet. Figure 28 illustrates how the supersonic bullet has developed its own shock-wave system in the quiescent atmospheric air as it outraces the decaying blast wave generated by

FIG. 23: A 500-TON HEMISPHERICAL EXPLOSION
A view of the 500-ton TNT charge (upper) used by DRB, Suffield, Canada, for tests
Prairie Flat (1968) and Dial Pack (1970). The bottom of the charge is supported by
styrofoam. The aerial photograph (lower) from Dial Pack was taken at 0 + 0.7
seconds. Note the natural shadowgram of shock wave S, against the backdrop of the
ground and the fireball. (Courtesy: Defence Research Board of Canada).

the explosive gas. Small fragments from the bullet also break through the blast wave to form their own shock-wave systems about the supersonic particles. The explosive high-pressure gas issues from the muzzle as a more-expanded turbulent supersonic jet, generating weak shock waves over the entire flow region encompassed by the blast wave. A person standing sufficiently close to the bullet's flight path would hear the crack produced by the shock system about the bullet, followed by the boom from the blast produced by the explosive gas that initially drove the bullet. Although the bullet generates head and tail shock waves, their separation distance is too small to be picked up by the ear as two distinct bangs. Were this not the case, *two* audible bangs would be heard rather than one. A separation in time of about 100 milliseconds (or over 100 feet in distance) is required before the ear can distinguish between these two waves (see Fig. 39). If one stands off the line of flight of the bullet, its sonic boom horseshoe pattern may have decayed to an inaudible level so that only the boom produced by the blast is heard.

FIG. 24: FIREBALL FROM A TNT EXPLOSION
The fireball from a 100-ton TNT explosion sometime after the formation of the blast wave. The smoke trails (Fig. 22) and dust raised by the explosion are also seen. (Courtesy: Defence Research Board of Canada).

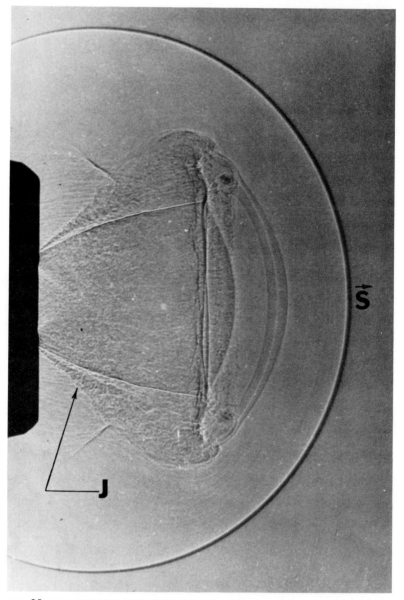

FIG. 25: SHOCK WAVES PRODUCED BY A BULLET

A series of photographs (Figs. 25 to 28) shows the emergence of a 0.30-inch-diameter bullet from a rifle barrel at 2200 feet per second (Mach number of 2), 75 microseconds after firing. The bullet is still in the barrel and pushes out a planar shock wave which then diffracts around the barrel and becomes a spherical shock wave, S. It is followed by an outflow (jet, J) of compressed supersonic air. (Courtesy: Ling–Temco–Vought, Inc.)

FIG. 26: SHOCK WAVES PRODUCED BY A BULLET

At 175 microseconds after firing, the 0.30-inch calibre bullet emerges from the muzzle driven by the explosive gases, E. (Courtesy: Ling–Temco–Vought, Inc.)

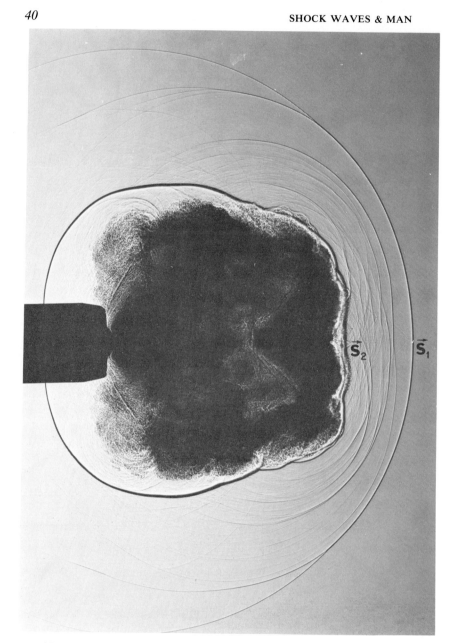

FIG. 27: SHOCK WAVES PRODUCED BY A BULLET
The explosive gases 250 microseconds after firing envelop the 0.30-inch-calibre bullet and drive a strong blast wave S_2, to overtake the earlier shock wave S_1. (Courtesy: Ling–Temco–Vought, Inc.)

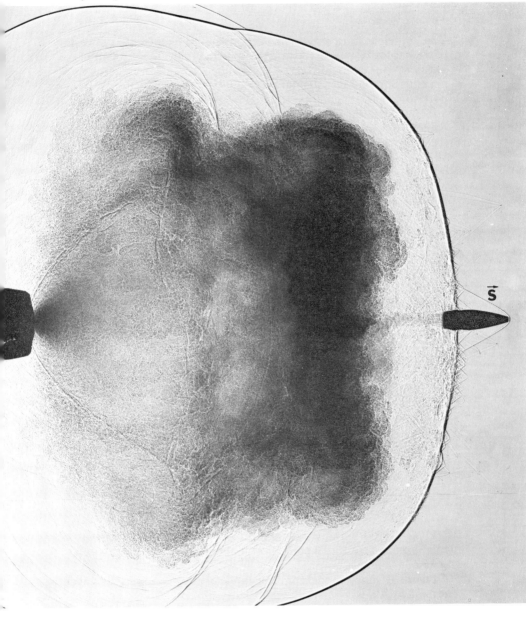

FIG. 28: SHOCK WAVES PRODUCED BY A BULLET

The 0.30-inch calibre bullet outraces the decaying blast wave and develops its own ballistic shock-wave patterns, 400 microseconds after firing. (Courtesy: Ling-Tem-co-Vought, Inc.)

Nuclear Weapons

The "efficiency" of explosive weapons can be expressed as the ratio of the explosive weight (destructive energy) to the total weight of the weapon. This definition ignores many important factors such as guidance, ease of delivery, initial cost, safety and security during transportation and storage, "kill effectiveness" (ability to destroy) against a target, and other considerations. For example, a simple chemical explosive bomb (which derives its energy from a chemical reaction), weighing 1000 pounds may carry less than 500 pounds of actual explosives, and it is then said to have an "efficiency" of 0.5 (pounds of explosive per pound of bomb). The first nuclear bombs of the type dropped on Hiroshima and Nagasaki had a yield equivalent to 20 kilotons of TNT (40 million pounds) packed into a bomb weighing about five tons (10 thousand pounds), thus giving an efficiency ratio of about 4000. The development of fusion explosives, and advances in weapon design over the following decade, improved the efficiency by another factor of 1000, yielding an overall efficiency of 4 million. Pound for pound, a thermonuclear weapon is nearly 10 million times more efficient than a chemical bomb.

It is worth noting that, while the detonation period is measured in milliseconds for a chemical bomb, it is a thousand times faster for a nuclear weapon (microseconds). In addition, the amount of uranium (U^{235}) or plutonium (Pu^{239}) required to give a "critical mass" for an explosive bomb is only about 110 pounds and 35 pounds, respectively. Approximately two pounds of U^{235} is no more than the size of a golf ball and could, if it was completely consumed in the explosion, release nearly 20 kilotons of TNT equivalent. Thus, the volume occupied by a nuclear explosive is very small compared with that required for a chemical explosive. The remaining nuclear bomb volume constitutes the hardware. Nuclear blast effects can therefore be analysed approximately as essentially due to a "point source" explosion, where a large amount of energy is concentrated in a very small volume. This concept will be discussed elsewhere. It is estimated that a megaton nuclear weapon, weighing about a ton, would require 10,000 railroad cars to carry its TNT equivalent of one million tons. Second degree burns would result from such an explosion over a 10-mile radius and a 50-percent probability of death from the blast would exist for a 5-mile radius.

Figure 29 shows models of three types of nuclear bombs displayed at the Los Alamos Scientific Laboratory Museum. In the centre is a gun-type uranium bomb (20 kiloton TNT equivalent, 10 feet long, 2.3 feet wide, weighing 9000 pounds), nicknamed "Little Boy", which was exploded over Hiroshima on 6 August, 1945; at the right is "Fat Man", an implosion-type plutonium bomb (20 kiloton TNT equivalent, 10.7 feet long, 5 feet wide, weighing 10,000 pounds) used on Nagasaki on 9 August, 1945; to the left is a thermonuclear device used for underground testing programs for peaceful applications of nuclear explosives in project Plowshare (named after Isaiah's vision of the world of tomorrow when "they shall beat their swords into plowshares"). It is worth noting that during World War II, the U.S.A. dropped, or otherwise set off, about 2000 kilotons of chemical explosives. Even the first atom bomb used on Hiroshima had an equivalent of 20 kilotons of TNT and today a single 50 megaton (50,000 kilotons) superbomb exceeds in power "all the chemical explosives used by the U.S.A. in World War II by a factor of 25."

It is beyond the scope of this book to detail the development of the atomic bomb. However, the very first test (Trinity), shown in Fig. 30, proved its tremendous destructive power. The mushroom-shaped cloud became a symbol of the cataclysm facing civilization.

The devastation that this 20 kiloton bomb brought to Hiroshima on 6 August, 1945, is illustrated in Figs. 31 and 32. Figure 31 shows one of

FIG. 29: NUCLEAR DEVICES FOR WAR AND PEACE
Nuclear bombs of the type used on Hiroshima (Little Boy – centre) and on Nagasaki (Fat Man – right) and a thermonuclear explosive device for project Plowshare (left) are shown in the Los Alamos Museum. (I.I. Glass, 1971).

FIG. 30: THE TRINITY ATOMIC TEST EXPLOSION

To prove the effectiveness of the first atomic device, after 28 months of intensive work by Los Alamos scientists, engineers, and technicians, a test was conducted, using the implosion type plutonium bomb, on July 16, 1945, in a desolate desert area (Jornada del Muerto) near Almogordo, New Mexico. The "unprecedented, magnificent, beautiful, stupendous, and terrifying" 20-kiloton blast (code name of Trinity) is shown above. Three weeks later Hiroshima and Nagasaki were obliterated. (Courtesy: U.S. Atomic Energy Commission).

44

FIG. 31: THE OBLITERATION OF HIROSHIMA

The first atomic bomb was exploded about 1800 feet above this Industrial Promotion Hall in Hiroshima at 8:15 a.m., 6 August, 1945. The steel and concrete ruin was one of the very few skeletons that remained and today it forms part of the Peace Memorial complex of Hiroshima. (Courtesy: City of Hiroshima Peace Memorial Museum).

the few shattered concrete and steel building shells, the Industrial Exhibition Hall (which was almost directly below the explosion) or the "Atomic Dome", whose skeleton remained standing after the bombing. This building forms one of the exhibits in the Peace Museum area of Hiroshima. Figure 32 illustrates the desolation following the blast and the remarkable rebuilding that has since taken

FIG. 32: A MODERN PHOENIX – HIROSHIMA

A view of the eastern part of Hiroshima immediately after the bombing (above) and at present (below). Hiroshima was rebuilt to be the most Western-looking city in Japan. (Courtesy: City of Himoshima Peace Memorial Museum).

place, to make Hiroshima the most Western-looking city in Japan. It is estimated that between 100,000 and 250,000 people perished from radiation, blast, fire and drownings. Radiation victims are being treated in a special hospital in Hiroshima to this day.

The shock wave produced by a shallow underwater nuclear explosion, made visible by the slick on the surface of the water, appears in Fig. 33. The condensation cloud is very striking, and its size can be judged by comparison to the naval ships. The structure of the entire phenomenon resembles Figs. 5 and 6, that is, the hot high-pressure expanding driver gas pushing before it a hemispherical shock wave. The photograph of another nuclear test explosion, Fig. 34, is a grim reminder of the symbolic atomic mushroom cloud. It was produced by a shallow underwater 20-kiloton nuclear test at Bikini in 1946. Owing to the high humidity and the decompression following the explosion, a condensation cloud about 6000 feet in width was formed. A spray dome due to the rupture of the gas bubble formed by the fireball (the hot, high-pressure core of the explosion), was generated before the condensation cloud. At its centre, the velocity of gases and vapours was about 2500 feet per second. These were vented through a column that went up to a height of about 6000 feet. The dome had a width of about 2000 feet and a wall thickness of approximately 300 feet.

Before the U.S.A., England and the U.S.S.R. agreed to stop poisoning our atmosphere with debris from nuclear weapons tests (France and China have still continued to do so), some high altitude tests were performed which are of scientific interest and are illustrated in Figs. 35 and 36. About 800 nuclear tests were performed between 1945 and 1970 by all powers. Figure 35 shows a schematic description of one high-altitude experiment. Johnston Island is about 1000 miles north of the geographic equator and Apia lies 1000 miles south of it, on the 170° West longitude line. Maui Island (in the Hawaiian group), which lies 800 miles north-east of Johnston Island, was the point of experimental observation. It can be seen that Apia and Johnston Island form magnetic conjugate image points; that is, electrons and ionized atoms produced at Johnston Island would gyrate around the magnetic field lines until they reached Apia, where they could be reflected back and forth between the conjugate points. On their journey, the electrons (with energies of 1 to 10 million electron volts) impact oxygen and nitrogen atoms and produce luminous excitation in the visible range, giving a red to green glow with an intensity ratio of about ten to one, respectively, thereby generating artificial aurora.

This phenomenon is well illustrated in Fig. 36 by four sequential pictures (they are very beautiful in the original colour). In the first picture, the entire sky is illuminated to daylight brilliance by the blinding flash of the explosion. As Maui Island is 70 miles below the horizon from Johnston Island, only the flash can be seen. In the next photograph, the shock wave, followed by the fireball, is quite evident.

The "aurora", as observed from Maui, is quite distinct in the last two photographs, as is the expanding shock wave and fireball containing the nuclear debris.

At an altitude of 190 miles, the shock wave moves at 5 miles per second (shock Mach number of 27). At a height of 310 miles, the shock wave only moves at about 2 miles per second. At this altitude, the Earth's magnetic field lines generate a counterpressure to the motion of the shock wave that is equal to the very low atmospheric

FIG. 33: AN UNDERWATER NUCLEAR EXPLOSION

The condensation cloud C, formed just after a shallow underwater nuclear explosion, and the slick S, due to the shock wave on the surface, are clearly illustrated. An appreciation of the tremendous size of the blast zone can be obtained by comparing it with the old destroyers and other naval vessels used in the test. (Courtesy: U.S. Atomic Energy Commission).

FIG. 34: THE MUSHROOM CLOUD

The formation of the hollow column in a shallow underwater nuclear explosion, whose mushroom top is surrounded by a late-stage condensation cloud, is shown in this awe- and fear-inspiring photograph. (Courtesy: U.S. Atomic Energy Commission).

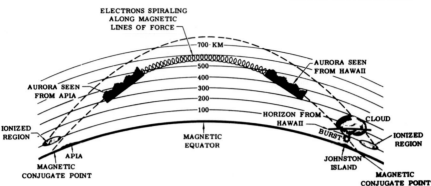

FIG. 35: PHYSICAL PHENOMENA GENERATED BY A HIGH-ALTITUDE NUCLEAR
EXPLOSION

Two magnetic conjugate (image) points 1000 miles north and south of the equator,
Johnston Island and Apia, respectively, were chosen for this test. A nuclear
explosion at high-altitude over Johnston Island generated high-energy electrons
shown spiraling along a magnetic field line and reflecting at its conjugate point
where on impact with oxygen and nitrogen they generated artificial aurora. (Courtesy: U.S. Atomic Energy Commission).

pressure. At still greater heights, this counterpressure slows the
shock wave even more, since it exceeds the ambient pressure.

The experiments verified analytical predictions about the production of artificial aurora, radio interference and other properties of
ionized gases and plasmas.

Providing ways can be found to contain the lethal debris safely, the
future possibilities for the peaceful use of nuclear explosives are
limited only by man's imagination. Yet, their invention has left man
and most living creatures precariously poised on the brink of possible
annihilation in a nuclear holocaust. The frightful consequences of
misusing them has brought man face to face with a crucial decision:
shall he be his "brother's keeper"? The potential for catastrophe has
prevented some major conflagrations in the last three decades. But it
has not prevented localized struggles, where millions of casualties
have occurred in various parts of the world since August, 1945. The
"smaller" wars, uprisings and fratricides, using conventional explosives, spears and knives, go on. It is worth noting that the U.S.A.
dropped more than three times the amount of explosives (6.2 megatons) on Indochina during the years from 1965 to 1971 than was used
in all of World War II (1941–45). No figures are available from the
other combatants.

The growth in stockpiling of nuclear weaponry, shown in Fig. 37,
is staggering. By 1960 there was a 10-ton TNT equivalent for every
man, woman and child on Earth. It may easily be one order of
magnitude (100 tons) greater in the Seventies. Included in this figure
are some quotations from the late President Kennedy and Senator

Pastore; it shows the development of the 50-megaton superbomb by the U.S.S.R. in 1961; and the years when atom and hydrogen bombs were exploded by the three superpowers, the U.S.A., the U.S.S.R. and China. The progressively decreasing time to develop the H-bomb from the A-bomb, especially by China, is startling. This unimaginable capacity for overkill would turn any nuclear war into a world holocaust. Undoubtedly, universal agreement to ban, disarm and disassemble all nuclear weapons as a first step, followed by a renunciation of conventional weapons, and ultimately leading to universal peace, is the most vital issue facing humanity in our time.

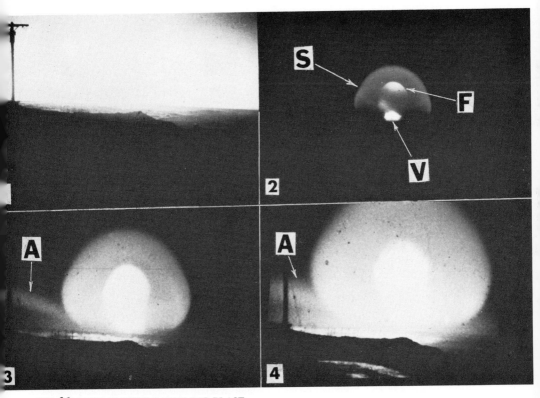

FIG. 36: HIGH-ALTITUDE NUCLEAR BLAST

A thermonuclear explosion 252,000 feet above Johnston Island was observed from Maui Island (Hawaii) 870 miles away. The blinding flash is shown in frame 1, and occurred at the instant of detonation (since Maui is 70 miles below the horizon from Johnston Island, only the brilliant flash but no shock wave is visible). Frame 2, taken 30 seconds later, shows the shock wave S, moving at 5 miles per second, the fireball F, and the swirling vortex V, containing the bomb debris. The artificially produced aurora A, at Hawaii is well illustrated in frames 3 and 4. The expansion of the entire phenomenon with time (see Figure 73) is also shown very graphically. (Courtesy: U.S. Atomic Energy Commission).

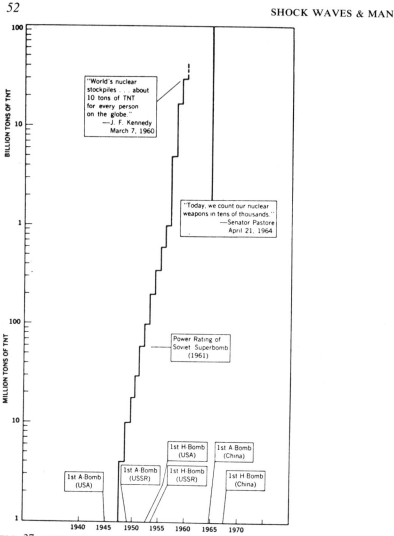

FIG. 37: OVERKILL

The growth of nuclear weapons from a few megatons in the late Forties to thousands of megatons in the Seventies, shows that the stockpiles provide a destructive power equivalent for every man, woman and child of hundreds of tons of TNT. Even a limited nuclear exchange would be a catastrophe. (Courtesy: R. E. Lapp).

Sonic Boom

With the advent of supersonic fighters and bombers in the 1960's and commercial supersonic transports (SST's) in the 1970's (British-French Concorde and Russian TU-144), a new era arrived. The SST's will transform a long, tiring journey from North America to Japan or

Australia into a relatively short, comfortable journey. It will also bring about the frequent exposure of man, creatures and structures to sonic boom.

Pressure levels with reference to the atmosphere (overpressure) produced during normal conversations are about one-millionth of an atmosphere or 0.002 pounds per square foot (psf). Sonic booms from present SST's are one thousand times greater or 2 psf. On the other hand, the eardrum ruptures at about 700 psf so that there is little danger of physical damage from sonic booms. However, the rapid rise in overpressure is very startling and jarring, and unless it is kept below 1 psf, it is experienced with annoyance by a significant fraction of people who have been exposed.

Sounds are transmitted at sea-level conditions (59°F) at 760 miles per hour. The speed of sound normally depends only on temperature and, at 65,000 feet, where it is -70°F, sound propagates at about 660 mph. When an aircraft travels at subsonic speed (Mach numbers less than unity, $M < 1$), the pressure waves travel in all directions (like the concentric wavelets generated by a pebble dropped into still water), and are too small to be heard. What we actually hear is jet or propeller noise (gliders are inaudible). As an aircraft approaches the transonic speed range (Mach numbers between 0.9 and 1.1, or $0.9 < M < 1.1$), or even surpasses it at supersonic ($M > 1$) and hypersonic ($M > 5$) speeds, the small pressure pulses pile up, ultimately to form two conical shock waves. The bow-shock has its apex at the head of the SST and the tail shock at the rear of the SST, as shown in Fig. 38. These waves strike the ground plane and are reflected back into the atmosphere. The entire space in front of the bow-shock is silent or undisturbed. The SST can therefore be seen and not heard, until some moments later, on arrival and reflection of the conical shock waves. When we do hear an SST, it could well be out of sight.

The picture is analogous to the surface waves and wake generated by a boat, and their reflection from the shore. The presence of the boat (although it is seen) is not felt until the disturbance wave hits the shore. Its magnitude depends on distance, boat size, and speed. Similarly, the strength of the boom (overpressure Δp) depends on the altitude of the SST, its lift distribution (which varies with angle of attack or inclination of wing to the airstream and how fast it is moving) and the slenderness profile (a stubby SST would cause strong booms).

As the conical shock waves encompass ever greater volumes of air away from the SST, their strength decays. The width λ, between the bow shock and the tail shock also increases slowly towards the ground such that the total impulse remains constant. A graph of overpressure Δp, over the wave length λ, has a sharp positive (maximum compression) rise at the bow shock (Fig. 38) followed by a uniform drop up to the tail shock, where it reaches a maximum negative value, immediately followed by a sudden recompression to atmosphere. The appearance of this plot in an ideal case looks like the letter N, and the boom is therefore also technically known as an N-wave. The sudden compressions at the bow and tail shocks give rise to the sharp startling cracks of the N-wave.

Some important features of the sonic boom are illustrated in Fig. 39. The upper portion of the figure shows how the conical bow and tail shocks intersect the ground to form a hyperbolic horseshoe-shaped pattern, which sweeps over everything in its path, making the boom felt by people, animals and structures in the boom corridor. The N-wave distribution (Δp and λ) around the horseshoe pattern are indicated schematically. It is seen that the maximum overpressure

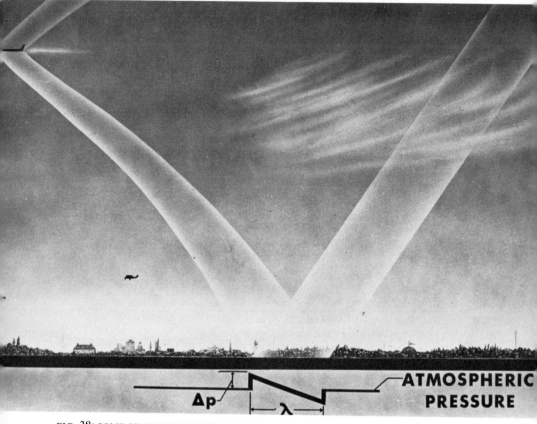

ATMOSPHERIC PRESSURE

FIG. 38: SOME CHARACTERISTICS OF SONIC BOOM
A schematic illustration of a sonic boom, and its reflection from the ground, as generated by a supersonic transport. The resulting pressure signature which looks like the letter N has a pressure Δp above and below the surrounding atmosphere of a few thousandth parts of an atmosphere (2–6 pounds per square foot), which rises abruptly (0.1–15 milliseconds), thereby causing a boom or bang. The duration of the N-wave, λ, varies between 100 and 350 milliseconds for a supersonic fighter and a large transport aircraft, respectively. (Courtesy: NASA).

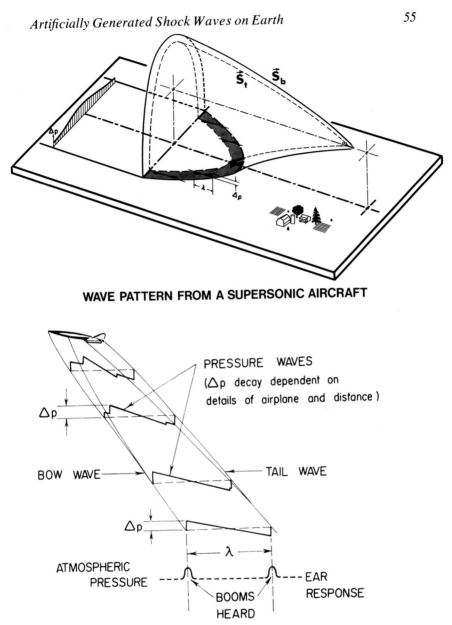

WAVE PATTERN FROM A SUPERSONIC AIRCRAFT

PRESSURE WAVES
(Δp decay dependent on
details of airplane and distance)

Δp

BOW WAVE — TAIL WAVE

Δp

λ

ATMOSPHERIC
PRESSURE — EAR
RESPONSE
BOOMS
HEARD

FIG. 39: ADDITIONAL PROPERITES OF SONIC BOOM

The upper sketch shows the horseshoe pattern formed by the intersection of the conical bow S_b and tail S_t shocks with the ground. Owing to refraction effects the audible widths of the horseshoe are limited to about 100 miles in the winter and 50 miles in the summer. The lower sketch shows how the complex shock wave system near the aircraft evolves and decays with distance into an N-wave of overpressure Δp and length λ. (Courtesy: UTIAS and NASA).

occurs directly below the SST and the minimum at the edges of the horseshoe, in a bell-like distribution (also sketched). The greatest width of this horseshoe, wherein the boom is still audible, depends upon the climatic condition of any given day. It may be as much as 100 miles on a cold day, or 50 miles on a hot day, for a supersonic craft flying at an altitude of about 50,000 feet (10 miles). It is worth noting that the ratio of corridor width to altitude is about 5. Beyond these widths, the waves are refracted back up into the atmosphere and do not reach the ground. A tail wind also increases the boom width and a head wind will decrease it. Basically the boom width increases with Mach number, tail wind and low temperature. All of these quantities reduce the effect of refraction to yield wider corridors.

The lower sketch of Fig. 39 shows the Δp curves, directly beneath the aircraft, as a function of altitude. It is seen that near the aircraft ("near field") the profiles are complicated by superimposed shock waves arising from engine nacelles, wings and tail plane. As one moves far away from the aircraft ("far field") these additional disturbances disappear and, as noted before, only an N-wave remains (the same is true of any type of explosion). The spreading of the bow and tail waves can also be seen. Finally, the ear response is also illustrated. It is seen that the ear responds mainly to the sharp rise-times of the N-wave signal. If the separation distance, λ, is of the order of 200 feet or greater (the speed of sound in air is about 1100 feet per second and at a Mach number of two the time interval would be 100 milliseconds), then two distinct booms are heard. If the interval is shorter, the ear cannot distinguish between the two pulses and only one boom is heard. (In the near field it might be possible to hear more than two booms if the delay times between the additional shocks are long enough).

For modern supersonic transports the N-wave duration is about 350 milliseconds and, for smaller fighter aircraft, about 100 milliseconds. The rise times at the bow and tail of the N-wave can vary considerably depending on factors such as atmospheric turbulence, temperature and wind. The turbulent eddies act like convergent or divergent lenses, and as a result the N-wave will have sharp spikes, which sound very startling or rounded, with a gentle sound. The degree of suddenness in the pressure rise to its maximum value (rise time) can be as little as 100 microseconds or as much as 15 milliseconds. The overpressures can range from a low of a fraction of a pound per square foot, up to several pounds per square foot. It is important to note that an overpressure (Δp) of two pounds per square foot is considered to be the upper acceptable limit for booms. About 40% of people tested rate this intensity as unacceptable; a 0.8 psf boom was considered acceptable; a 3.6 psf boom was judged completely unacceptable. During the early years of supersonic flying, some military pilots had inadvertently flown their fighter aircraft at altitudes which were too low to allow the shock waves to decay to acceptable levels of about 2 psf. As a result hundreds of window were smashed with damage running into tens of thousands of dollars.

An overpressure of 2 psf can readily be achieved by running up or down three flights of steps (about 30 feet). However, the rise time is so gentle that the ear is unaffected. (In an elevator the rate is increased and we feel the pressure changes). Therefore, production of a sinusoidal overpressure distribution would be ideal for eliminating the sonic boom. Unfortunately, even distributions such as these generated by an aircraft, would naturally steepen to give sharp-fronted shock waves with short rise times, thereby generating the startling crack which people find so objectionable. The degree of annoyance depends on the sensitivity, age, health, and other psychological, sociological, and physiological states of an individual at the moment the boom is experienced. This subjective individual response makes it extremely difficult to assess quantitatively the effects of sonic booms on human beings.

The shadowgram in Fig. 40 shows the decay, stretching, and reflection of the bow and tail shock waves generated by a supersonic spherical projectile. Figure 41 illustrates the "near field" about two wing-body models in a supersonic wind tunnel test section at a flow Mach number of 2.5 The schlieren photographs illustrate the wave systems about blunt and sharp nosed bodies. The bow waves, and the shock waves due to the wing, are readily seen. These two figures were included to show that the previous schematic sketches are represen-

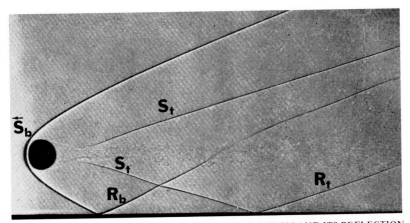

FIG. 40: THE SHADOWGRAM OF AN ACTUAL SONIC BOOM AND ITS REFLECTION
The bow (S_b) and tail (S_t) shock waves generated by a supersonic spherical projectile and their reflections (R_b, R_t) from a plane solid surface simulate some of the characteristics of the N-wave sketched in Figs. 38 and 39. They clearly show the spreading of the N-wave with distance before and after reflection. Unlike the SST, the sphere has no protuberances to disturb the N-wave pattern. (Courtesy: *Selected Topics on Ballistics*, edited by W. C. Nelson, AGARDOgraph 32, Pergamon Press, New York, 1959).

tative of actual flows. Figure 42 extends the illustration to hypersonic Mach numbers and high altitude or low-atmospheric density where the bow wave is still clearly visible but the details of other, weaker shocks are no longer recorded. Figure 43 shows the "near field" boom signatures taken at 1994 feet above, and 1654 feet below, a bomber aircraft flying at an altitude of 45,000 feet at a Mach number of 1.65. The shock waves produced by the fuselage bow, nacelles, wing and tail are quite discernible in the overpressure signatures. It should be noted that at 45,000 feet the atmospheric pressure is only about one tenth that at sea level, say 200 psf. Consequently, a pressure rise of 3.75 psf is a very large fraction (0.02) of the local atmosphere, compared to a sonic boom on the ground of 2 psf. This latter fraction, 0.001 of an atmosphere, is one order less.

Eight (a to h) important parameters that affect the overpressure, Δp, generated by a sonic boom appear in Fig. 44. It can be seen that if all other factors are held constant as each variable is examined, the heavier aircraft (a) in flight will create the larger boom. (A modern SST such as the Concorde has a length of 204 feet, a maximum take-off

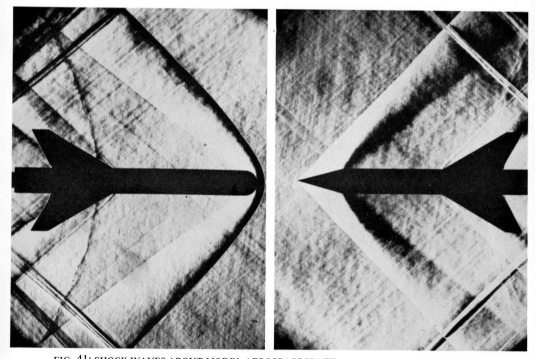

FIG. 41: SHOCK WAVES ABOUT MODEL AEROSPACECRAFT
Schlieren photographs of the wave systems generated about blunt-nosed and sharp-nosed supersonic models at a Mach number M = 2.5 in the UTIAS 16 x 16 inch supersonic wind tunnel. (Courtesy: UTIAS).

weight of 367,000 pounds, and with 132 passengers, carries a maximum payload of 25,000 pounds up to 4220 miles at 64,000-foot cruise
altitude at a Mach number of 2.2). The overpressure on the ground
decreases with increasing flight altitude (b) but there are height limits
imposed by engine performance and lift (and possible hazards from
solar flare radiation). The overpressure increases with greater cross-
sectional area (c) as stronger shock waves are generated for stubbier
bodies. (For the same cross-section the overpressure decreases as the
length of the aircraft is increased to make a more slender body). As
can be seen, the effect of speed or Mach number is not particularly
significant (d). This is due to an interplay between speed and angle of

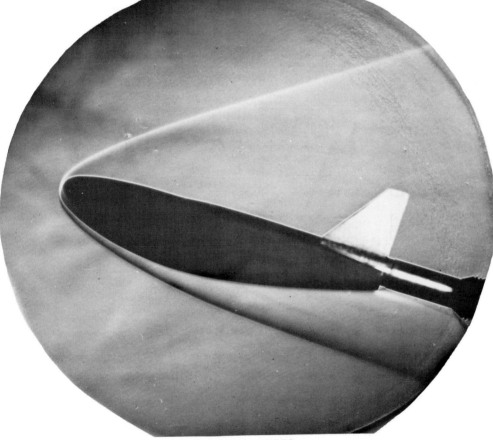

FIG. 42: SHOCK WAVES AT HYPERSONIC SPEEDS
A laser-schlieren photograph of the wave system about a hypersonic aircraft model
in the 16 × 16 × 24 inch long open-jet test section of the UTIAS Hypersonic Shock
Tunnel Mark II, at a Mach number M = 8.2, where the stagnation temperature at
the blunt nose was approximately 5000°C. (Courtesy: UTIAS).

attack to provide a given lift. However, a small decrease with higher Mach number is predicted. Changing weather conditions (e) during the day such as air pressure, temperature, humidity and wind, will all affect the refraction of the shock wave, with consequent important variations in overpressure. Atmospheric turbulence (f) will modify the N-wave to make it more spikey, with several times the nominal overpressure (superbooms), or more rounded. As noted above, these

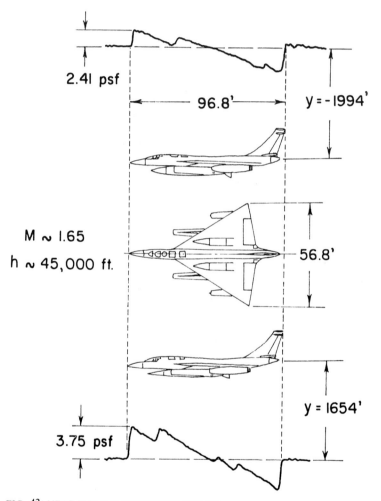

FIG. 43: NEAR FIELD N-WAVE SIGNATURES
Actual sonic-boom pressure signatures recorded 1994 feet above and 1654 feet below a supersonic aircraft flying at an altitude of 45,000 feet and Mach number M = 1.65, showing bow and tail shock waves and additional shock waves from various parts of the aircraft. The nearly 100-foot-long aircraft is shown in side and plan views to indicate the origin of near-field shock waves from the engine nacelles and the tail section. (Courtesy: NASA).

conditions result in sharper (traumatic) or more muffled (gentle) sonic booms. The rest of the N-wave signal in both cases is also made more noisy by turbulence. It is important to note that a 0.7 psf overpressure with a 1 millisecond rise time sounds just as loud as a 2.5 psf boom with a 10 millisecond rise time. However, the potential structural damage from the latter boom would be much greater. Therefore, both are important when taking loudness effects into account. Finally, as a

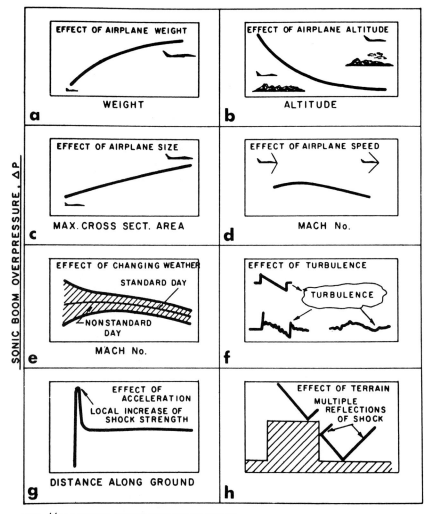

SONIC BOOM OVERPRESSURE, ΔP

a — EFFECT OF AIRPLANE WEIGHT — WEIGHT

b — EFFECT OF AIRPLANE ALTITUDE — ALTITUDE

c — EFFECT OF AIRPLANE SIZE — MAX. CROSS SECT. AREA

d — EFFECT OF AIRPLANE SPEED — MACH No.

e — EFFECT OF CHANGING WEATHER — STANDARD DAY — NONSTANDARD DAY — MACH No.

f — EFFECT OF TURBULENCE — TURBULENCE

g — EFFECT OF ACCELERATION — LOCAL INCREASE OF SHOCK STRENGTH — DISTANCE ALONG GROUND

h — EFFECT OF TERRAIN — MULTIPLE REFLECTIONS OF SHOCK

FIG. 44: FACTORS AFFECTING SONIC BOOM

Schematic illustrations sketch the effects on sonic-boom overpressure Δp, resulting from SST weight, altitude, length and cross-section, speed, daily weather changes, air turbulence, accelerating maneuvers, and reflections from terrain and buildings. (Courtesy: Boeing Aircraft Co.).

result of turbulence, the same boom for two observers at two different points can be acceptable to one and intolerable to the other. Also, during maneuvres, an accelerating aircraft (g) can cause focussed booms or superbooms (6 to 15 psf) that can be several factors above the tolerable 2 psf overpressure maximum, with a consequent sharp increase in startle and annoyance. The reflection of a sound pulse from a plane surface causes a doubling in overpressure (h). Inside a concave corner, this can increase tenfold due to multiple reflections. Therefore, multiple reflections caused by buildings or terrain topography can increase the overpressure of sonic booms and must be considered in any analysis of their effects on the people and structures of a community. Even the triggering of a mountain avalanche by a sonic boom might be a remote possibility.

Many people already have been subjected to sonic-boom tests by supersonic aircraft to determine community response. Whether most people will find the boom from present-day sst's (Concorde, TU-144) acceptable is still an open question. Consideration will also have to be given to wildlife and to heritage buildings, all of which are subject to possible damaging effects from the sonic boom. It is not known if herds of reindeer or other animals will stampede; or if birds will be startled sufficiently to disrupt nesting habits with a resulting danger to survival of the species; or if cherished heritage buildings, built by famous architects and with a long history behind them will receive irreparable damage from sonic booms accelerating the natural aging process caused over the years by temperature and humidity cycling, rain, snow, sun and vibrations. Numerous damage claims have already been made by people whose houses and businesses have suffered from sonic booms. Considerable sums have already been paid as settlements both in and out of court. Much data must still be gathered to formulate a complete picture of the effects of sonic boom on human physiological and psychoacoustic responses, as well as the responses of wildlife and structures. Considerable research is being continued in these areas in Britain, France, u.s.s.r., u.s.a. and Canada. The builders of present-day sst's are confident that they will be accepted by the public and that the second generation sst's, which will have a low-enough overpressure (less than 1 psf), should have no difficulty on that score.

Physiological effects on humans, such as hearing impairment, are not evident. However the startle effect and interruption of rest, sleep, and concentration and upon conversation, music, tv and radio are obvious enough. Some adaptation appears possible (very much as one gets used to thunderstorms, which are many times more severe than man-made booms, although relatively infrequent). However, the random superbooms from atmospheric turbulence effects and maneuvers may make the noise less bearable. As Ribner (1972) points out, "*individual* human response to an *individual* boom is highly variable because of personal and sociological factors: hence it is not very predictable''. These factors tend to average out when a group of people are subjected to an individual boom, and even more so to a collection of booms. Consequently, the effects of booms become more amenable to prediction from community-survey data. This may be compared to accident prediction. We cannot foresee an individual road accident but the *total number* that will occur in a given holiday week-end can be predicted with reasonable accuracy.

Atmospheric Re-entry Phenomena

With the launching of Vostok 1 on 12 April, 1961, cosmonaut Yuri Gagarin opened the age of manned-spacecraft explorations. He circled the globe at 5 miles per second, in a vehicle weighing 10,500 pounds, in 1 hour and 48 minutes, reaching a maximum altitude of 203 miles. At such a velocity, each pound of mass in orbit has an energy equivalent of seven pounds of TNT. Consequently, the re-entry of a space capsule into the denser layers of our atmosphere is just as spectacular as a meteoroid entry or an explosion.

This process is well-illustrated in Fig. 45. The sketch follows John Glenn's description of the first American orbital re-entry in his Mercury-Atlas 6, weighing 3000 pounds, on 20 February, 1962, after 3 orbits lasting 4 hours and 55 minutes, and reaching a maximum height of 162 miles. One can see that the entire capsule is enveloped in a spectacular plume of fire. In front of this plume there is a glowing bow shock wave, its shape modified by the retropack strapped to the heat shield, that caused some anxiety on re-entry. A space-capsule model (without retrorockets) test of the same phenomenon in a hypersonic shock tunnel appears in Fig. 46. The glowing gas cap and bow shock are well illustrated.

About 99% of the energy expended by the plunging capsule is dissipated through the shock wave by compressing and heating the gas flowing around it at temperatures up to 6500°C—hotter than the Sun's surface temperature of 5700°C. The oncoming atoms and molecules striking the gas envelope around the craft at 5 miles per second give up their directed energy to produce thermal energy. This thermal energy is divided up (partitioned) among the various states of motion of molecules and atoms, such as motion along a path (translation), rotation, molecular vibration, tearing apart of molecules (dissociation) or atoms (ionization). The partition processes among all the states (modes) are achieved after hundreds or thousands of molecular collisions through the shock front that may be many inches thick (relaxation distance) at altitudes of 300,000 feet, and only a fraction of an inch at 50,000 feet, where the ambient pressures are 10^{-6} and 10^{-1} of an atmosphere, respectively, and the space capsule Mach number has decreased considerably.

Most of the heat, then, is dissipated in the wake and the rest is transferred to the body by conduction through the viscous boundary layer on the heat shield. Heating by radiation plays an increasingly greater role as the re-entry velocity increases. From the Moon, this re-entry velocity is 6.8 miles per second and from Mars, about 8.1 miles per second. The heat shield thickness for lunar re-entries is about 6 inches. This ablating, charring, frothing and gassifying shield is made of fiberglass, polymers, epoxy and phenolic resins, or other composite plastic materials. It is fortunate that the bow shock wave dissipates nearly all of the energy of motion of the capsule while slowing it down sufficiently for a parachute landing, otherwise the heat shield would have to be inordinately thick and heavy. If this were

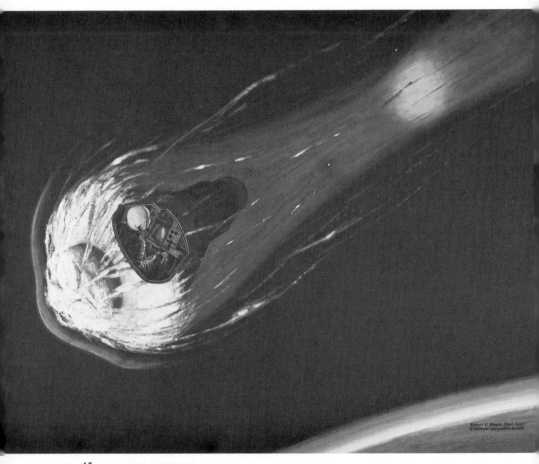

FIG. 45: MAN IN A METEOR

John Glenn, first American to orbit the Earth on 20 February, 1962 had an anxious and dramatic re-entry. It was thought that the heat shield was in danger. In actual fact it was only the straps holding the retrorocket pack that burned away. He had a safe re-entry. (Courtesy: © National Geographic Society).

the case, it is doubtful if some of the flights could have been made, for more weight would have gone into the shield at the expense of the crew, their life support systems, and instrumentation.

As the vehicle slows down to subsonic velocities, the bow and tail shock waves surrounding it race ahead to reach the ground before the capsule. A shock wave, once formed, must dissipate itself like a blast wave from any explosion. Consequently, sonic booms are heard on

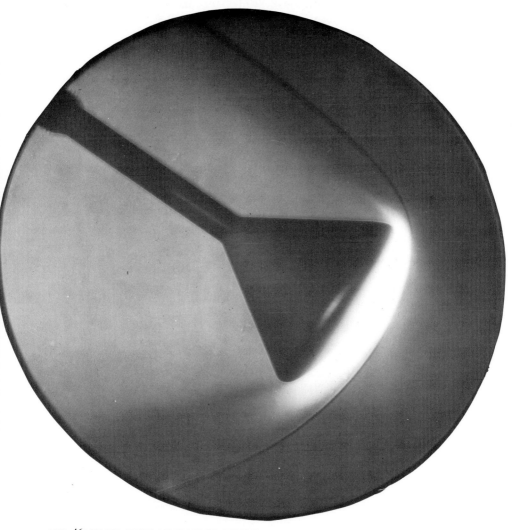

FIG. 46: SIMULATION OF SPACE RE-ENTRY
A laser-schlieren photograph of a re-entering space capsule model at a Mach number M = 10.2 in the UTIAS Shock Tunnel Mk II showing the bow shock wave and glowing air nose cap at a temperature of about 5000°C. (Courtesy: UTIAS).

the ground or on board recovery ships at sea before the spacecraft touches down. The television and radio commentators covering such a re-entry splash-down have often made reference to these sonic booms during their broadcasts.

Some recent sonic boom data obtained from the re-entry of Apollo 15 show overpressures of about 0.4 psf as measured on the recovery ship USS Kawashiwi, which corresponds to a generating point along the flight path at an altitude of 110,000 feet and a capsule flight Mach number $M = 4.6$. An overpressure of about 0.8 psf on board USS Okinawa corresponded to a similar point on the trajectory at an altitude of about 20,000 feet and a flight Mach number of 1.2. A boom overpressure of about 0.2 psf was also measured on board the recovery ship USS Genesee. It was estimated that this value arose from a point at a flight Mach number of 15.7. Consequently, the boom signature for this decelerating maneuver reached a maximum of about 1 psf at the lower altitudes and Mach numbers. This value is much less than the boom overpressure for current supersonic transports and can be attributed to the insignificant lift that is generated by the capsule (compared to a wing, for example) and its small volume compared to an SST. Although the blunt base of the conical capsule generates a strong re-entry shock wave in the near field, by the time the boom reaches the ground its influence is not effective.

4

SHOCK WAVES IN SPACE

The last two chapters provided ample evidence that shock waves generated on Earth from natural causes, or by man, can have disastrous effects. In space, however, shock-wave phenomena, which occur incessantly among the countless stars, nebulae and galaxies, can be observed by men with a sense of awe, wonder and excitement but without concern for possible cataclysmic effects on Earth.

The great masses of gas in our Sun and in the stars are heated by nuclear reactions to interior temperatures of hundreds of millions of degrees at enormous pressures and incredibly high densities. Exploding shock waves, generated from the rapid heat addition of the nuclear reactions, can move from the dense interior to the rarefied gas at the outer surface of a star. There a shock wave can reflect back as an *implosion* (reversed explosion) wave. The process can repeat and give rise to pulsations. Similar results can arise from an interplay between pressure and gravitational forces. Cosmologists currently believe that even the creation of our entire universe may be modelled after such a cyclic explosion-implosion process. However, this new discipline of cosmic gasdynamics is still in its infancy and undoubtedly many modifications to current simple theories will be developed in the future to accommodate new facts about the universe as they are learned from space explorations and other sources.

Solar Wind

The solar furnace radiates energy at the stupendous rate of 4×10^{26} watts, and since a megaton bomb has an energy equivalent of 4×10^{15} watt-seconds, therefore the sun every second emits energy with a TNT equivalent of 10^{11} megaton bomb blasts. Of this, the Earth receives the energy equivalent of a 50 megaton bomb blast per second. This is quite sufficient to supply the energies required to generate thunderstorms, winds and other thermal phenomena on Earth.

The Sun also issues a steady "solar wind" stream of fully-ionized particles (all electrons are removed from their shells) consisting of electrons, and nucleons of hydrogen, helium and some heavier elements. This plasma is electrically neutral (uncharged). The solar wind "blows" with a supersonic velocity of about 250 miles per second with a bulk number density of about 160 particles per cubic inch at an average kinetic temperature of about 100,000°C. The solar wind is able to drag with it some of the magnetic field lines from the sun and the system of particles and fields interacts with those of the planets, their moons, comets and meteors.

The interaction of the solar wind with the Earth and its magnetic field is sketched in Fig. 47. Basically, the dipole magnetic field surrounding the Earth is reshaped by the magnetic pressures arising from its interaction with the solar wind particles. A streamline shape results, which is called the magnetopause. It may be pictured as the boundary line where further penetrations of the ionized particles into the Earth's magnetic field are prevented by forces developed as the particles cut the field lines. In the interior of the streamlined magnetopause is the Earth and its Van Allen radiation belts. Exterior to it is a collisonless bow and tail shock-wave system as if the mag-

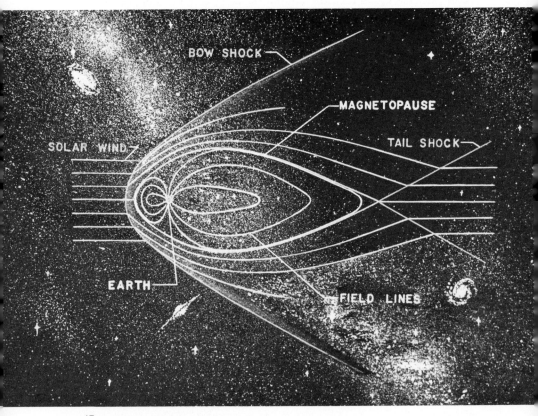

FIG. 47: MAGNETOSPHERE FLOW
The shock waves in space generated by the interaction of the solar wind with the Earth's magnetic field. (Courtesy: UTIAS, after Levy et al. 1964).

netopause were a model in a hypersonic wind tunnel at a flow Mach number of about 10 (see Fig. 46). The stagnation point (where a moving particle would come to rest at the front centre of symmetry of the magnetopause) is about 20 Earth radii (80,000 miles) from the Earth's centre; the bow shock about 30 Earth radii.

The shock thickness under these rarefied conditions has an incredible value of some thousands of miles. Alternatively, the thickness is of the order of a few ion cyclotron radii, that is, the radius of gyration of an ion about a magnetic field line. As the mean-free-path of a particle approaches infinity at these rarefied conditions the shock thickness can no longer be defined in its terms, since it would be meaningless. As a matter of fact, the changes in properties across the shock wave are no longer due to collisions. That is, the existence of the magnetopause is not "transmitted" to the shock wave by particle collisions as is usually the case in a denser gas but by electromagnetic interactions. It is therefore called a collisionless shock wave. The subsonic plasma behind the shock wave is heated by the interactions to temperatures at least tenfold greater than in the solar wind in front of the wave.

Although Fig. 47 gives a general idea of this type of flow, a situation that has been verified by data from scientific satellites as far as bow shock and magnetopause shapes are concerned, the details are much more complex and also time-dependent, as the solar-wind flow depends on the activity of the sun. For example, Fig. 48 gives additional information on some current ideas regarding magnetosphere flow.

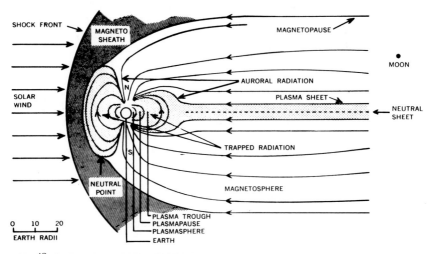

FIG. 48: DETAILS OF MAGNETOSPHERE FLOW

The "doughnut and tail" model of the Earth's magnetic field, representing present concepts after a decade of observations in space. The dot marked "Moon" indicates the relative distance at which the Moon's orbit intersects the plane of view. (Courtesy: Engineering, Cornell Quarterly).

The magnetopause cavity resembles that of a comet, with a long tail extending up to several million miles downwind. The position of the Moon is also shown, but since it has no atmosphere or magnetic field, the solar wind particles are simply absorbed by the lunar surface without creating a shock wave system. The open and closed magnetic field lines define a number of regions. The neutral points divide the magnetic field lines that close on the Earth and those extending to the tail. The neutral sheet separates the open magnetic field lines directed towards the Earth and connected to the North Pole from those coming from the South Pole.

A flow of plasma takes place from the Earth, along the open magnetic field lines, into the tail. This supersonic jet is known as the polar wind by analogy with the solar wind. The solar plasma flows mainly around the magnetopause. There is evidence, however, that some of it enters the Earth's atmosphere in the polar regions through the neutral points and through the tail, replenishing the Van Allen belts, while at the same time generating brilliant aurora nearer the Earth's atmosphere and causing radio blackouts. The processes involved are still not well understood. Comets in their transit near the Sun may well develop similar flows.

Solar Flares

The Sun, whose diameter is nearly twice as great as the diameter of the Moon's orbit about the Earth, is a seething mass of hot ionized gases intercoupled with electromagnetic fields. The resulting motions are very complex and not well known. They are further complicated by surface explosions fed by magnetic-field energy near sunspot activity that appear to give rise to solar flares. Figure 49 shows such a flare shooting out 110,000 miles into space, at a speed of about 930 miles a second. The ionized-gas particles are frozen into the magnetic lines like beads on a string, giving rise to the various loops on the photograph. The enormous energies involved in electromagnetic radiation and particle motion in some flares may be of the order of billions of megaton bombs. Many flares extend outward farther than that shown in Fig. 49 and some of them may, in fact, be represented by the model shown in Fig. 50.

When a strong flare occurs, electromagnetic radiation in the form of ultraviolet, visible, X-rays and radio waves travel outward at the speed of light and reach the Earth in about eight minutes. Some cosmic rays (highly-energetic, fully-ionized nuclei or nucleons of hydrogen, helium and other elements) may reach the Earth slightly later. However, as shown in Fig. 50, the bulk of the ionized gas, at temperatures of about a million degrees centigrade locked into the magnetic field lines of the Sun, act as a piston moving at about 620 miles per second. This expanding, blunt, tear-drop piston generates a strong collisionless bow shock many thousands of miles thick. On arrival at the Earth, the shock wave and ionized gases interact with the Planet's magnetic field. They envelop and shield the Earth from galactic cosmic rays, which may be more energetic than the solar cosmic rays. The galactic cosmic rays that get through the stretched solar magnetic shield excite Alfvén waves, (named after the Norwegian Nobel Laureate H. Alfvén) much like plucked strings. Although the temperature and pressure ratios across the bow shock are significant, the gas is so tenuous that it hardly changes the physical state of our atmosphere. However, the ionized gas profoundly affects radio communication and the generation of aurora. The possibility also exists of exposing astronauts and future passengers of high-altitude hypersonic transports to an unacceptable radiation dosage. This might be especially hazardous, despite shielding by the magnetosphere, in the high-energy penetration range whether from electromagnetic radiation such as X-rays, or from fast-moving nucleons or electrons.

FIG. 49: SOLAR FLARE
The loop prominence produced by the intense magnetic field in an explosive flaring solar region extends about 110,000 miles into space.
(Courtesy: U.S. Air Force).

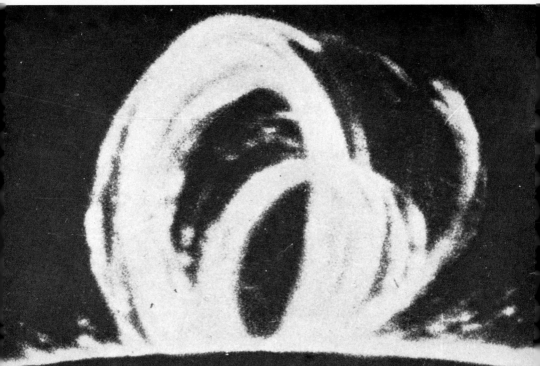

A recent massive explosion generated a huge solar flare which was observed and investigated by the astronauts of Skylab 2, on 5 September, 1973. There was a 20-minute period, after the flare reached its maximum, when short-wave radio communication over large areas of the Earth faded out completely. Such a flare would be hazardous to the crew of a lunar mission but was not considered harmful to the Skylab crew, owing to the magnetic shielding provided by the magnetosphere. It was estimated that the energy of the explosion would be sufficient to supply the world's total energy needs for the next 500 years. At the present use of about 10^{15} BTU/day, this is equivalent to nearly 50,000,000 megaton bombs! As noted above, some flares can be several *thousandfold* larger.

It is difficult to predict when a solar flare will occur. Every sunspot

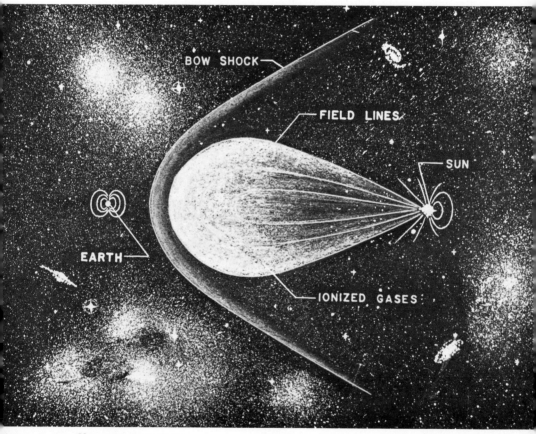

FIG. 50: SOLAR EXPLOSION

A shock wave in space generated by a solar eruption. The sketch shows the fully ionized nucleons attached to the solar magnetic field lines acting as the driving piston for the shock wave. (Courtesy: UTIAS, after Gold, 1962).

complex can be considered as a potentially dangerous eruption source, likely to impact the Earth from two days before, to seven days after, the central meridian has passed. Similar events are observed in so called flare-stars. An understanding of this process in our own Sun will undoubtedly provide the knowledge to interpret similar phenomena in other parts of the universe that are still mystifying astronomers and astrophysicists.

It is also worth noting that the corona, the tenuous outer layer of the Sun, has an asymmetric glow during a sun-spot minimum (Fig. 51) and radiates symmetrically (as seen during a solar eclipse) during maximum activity. It is believed that shock-wave acceleration and heating may generate the symmetry, as shown in Fig. 51.

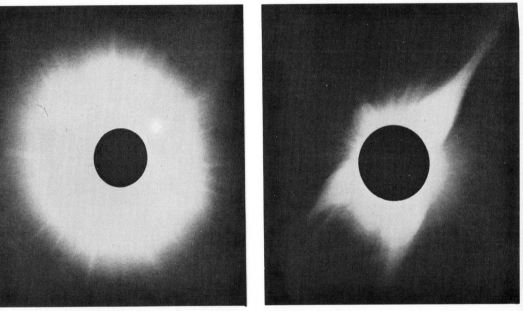

FIG. 51: SHOCK WAVE HEATING OF THE SOLAR CORONA
The greatly different shapes assumed by the corona at a sun-spot maximum (1947, left) and minimum (1952, right) are clearly illustrated. The differences are attributed to shock-wave heating, making the corona more symmetrical. The plasma temperature in the corona is approximately 1,000,000°C. The Sun's surface temperature is about 6000°C. (Courtesy: Yerkes Observatory).

Exploding Stars and Galaxies

FIG. 52: STELLAR EXPLOSION

The Crab nebula in Taurus, 4400 light-years away, is expanding at about 800 miles per second. A burst of light marking the birth of the supernova, which originated about 3400 BC, was first observed in 1054 AD by the Chinese observatory in Peking. (Courtesy: California Institute of Technology).

Most stars have shown little change over the years of observation so far as one can tell with the limited equipment of today. Occasionally however, a bright star suddenly appears, where no object was previously discernible. Such a new star has been called a nova, or in the case of exceptional brilliance, a supernova. Within a few hours, its brightness can go up to fifty thousand times its original value for a nova and perhaps a billion-fold for a supernova. It is believed that such events are caused by explosions of truly cosmic dimensions.

An example of this type of explosion is seen in Fig. 52. The Crab nebula (the ragged filaments are reminiscent of the legs of a crab), in the constellation of Taurus, is the remains of a supernova that was recorded in the astronomical observatory of Peking, China, in 1054. As this nebula is about 4400 light-years away from Earth, the actual explosion took place about 3400 BC. The gases are still expanding at a rate of about 800 miles per second. It now has a major-axis diameter of approximately six light-years, about 70 per cent greater than the distance from Earth to the nearest star, Alpha Centauri (25×10^{12} miles). It is estimated that the enormous energy release was equivalent to 25×10^{38} megaton bombs or the Sun's radiant energy for 10^{10} years. This illustration serves to emphasize the astronomical dimensions involved in a cosmic event.

The appearance of the Crab nebula is not unlike that of the laboratory exploding glass-sphere experiment (Fig. 5) showing the expanding high-pressure gas after the explosion. In the nebula, the expanding gases would drive a cosmically-thick, collisionless shock wave rather than the thin shock wave produced in the laboratory. The interaction of the ionized gases with its magnetic field gives rise to electromagnetic radiation over a wide frequency range. This nebula is well known as a gamma ray, X-ray and radio source. It is believed that a pulsar or neutron star exists in its interior.

The present model for this type of explosion assumes that, when the thermonuclear reactions in a star approach exhaustion (with iron accumulating in the inner core), the gas pressures become weaker and the star contracts or collapses, due to gravitational forces, to become a tiny white dwarf, thereby generating an implosion that subsequently reflects from the point of collapse in the interior as an explosion. A considerable amount of the mass of the star is thrown into outer space by the explosion as in the case of the Crab nebula. This gas contains all the elements up to uranium, the most complex element occurring in quantity on Earth. Elements heavier than uranium must have broken down so that they are no longer detectable on our planet. It is instructive to speculate on the possibility that our Solar system formed from just such an accumulation of explosively ejected gas with the planets coalescing from the outer material. Consequently, the Earth, being too small and too hot at its birth to hold hydrogen and helium, ended up with heavier atoms. Its core is thought to be liquid iron that may have originated from the exploding inner core of the supernova. Finally, we may even conjecture that life on Earth may well owe its origin to an explosion (see the closing remarks of the next subsection). It is presently thought that the remainder of the material from an exploding star can collapse further to form a stable neutron star. The density becomes incredibly great (100,000,000 tons per cubic inch). Further collapse is possible to a radius where emitted or incident photons (light) are captured by the gravitational field and are unable to escape, giving rise to a so-called "black hole" which is still far from being understood. Its radius is also called an "event horizon" by astrophysicists. It is readily determined from the mass of a star and the speed of light by equating the centrifugal force on the photon to its gravitational attraction and solving for the radius. The photon is then analogous to a circular orbiting satellite.

Explosions occur not only in stars but also in galaxies, on a scale enormoulsy greater than a supernova. Millman has noted that atoms, men, stars and galaxies may be used as qualitative units to obtain an idea of the variation of size in the universe. Each unit goes up by orders of 10^{10}; that is, 10^{10} atoms stretched in a row yields the height

FIG. 53: GALACTIC EXPLOSION

This tremendous explosion in the central region of the galaxy M-82 occurred about 12 million years ago. Such explosions can account for the very energetic galactic cosmic rays and radio waves. The explosion energy was calculated as the equivalent of 5×10^{42} megaton bombs. (Courtesy: Hale Observatories).

of a man, and so on. Consequently, we can expect scales of explosions in galaxies to be increased by many orders of magnitude. Such an explosion is shown in Fig 53. It is estimated that the explosion in the central region of galaxy M-82 took place about 12 million years ago with an energy equivalent to 5×10^{42} megaton bombs, or some ten thousand times greater than for the Crab nebula. The energies are usually estimated by using classical nuclear or point-source-explosion theory but this method might be in error by a few orders of magnitude. These enormous energies can account for the production of galactic cosmic rays and other electromagnetic radiation observed on Earth and in space.

Cosmological Big Bang

Some cosmologists believe that our universe came into being about 15 eons (billion years) ago with a big bang. At that time (zero-time) all the matter and energy in the universe was squashed together in one huge bomb (nuclear or cosmic egg) and, owing to some instability, this primordial bomb suddenly exploded. It would certainly have produced the most stupendous explosion of all time. The fragments of this gigantic explosion became the galaxies and, like the glass fragments shown in Figs. 5 and 6, were sent hurtling in all directions and are continuing to do so to this day. This theory originated with Lemaitre in 1927, and was later extended by Gamow, who called it the "big-bang theory" or the "exploding-universe theory". This theory, presented in 1948, in collaboration with Bethe and Alpher is more sophisticated and explains many possible processes on an atomic scale.

The universe is now essentially composed of about 90 percent hydrogen and 9 percent helium, with the remaining 1 percent accounting for the more complex atoms. The heavier atoms result from fuelling the astral thermonuclear furnaces in the universe. Consequently, then, the nuclear egg must have consisted of hydrogen squeezed together to form neutrons with a density of 100 million tons per cubic inch, as for a neutron star. However, on exploding, the cosmic egg disintegrated with astounding ferocity into separate neutrons which broke down into the primary atomic building blocks, namely protons and electrons. In this manner, through the process of collision, the stable atoms as we know them today were built up starting with the hydrogen proton. As the temperatures fell, the various nuclei attracted electrons to form neutral atoms. A great agglomeration of atoms could condense to form galaxies and stars as they were speeding away from the source of the cosmic explosion. There are some difficulties with this simple model of trying to build more complex atoms beyond helium that has led to the formulation of the so-called, "steady-state theory," by Bondi, Gold and Hoyle, who stipulate that simple hydrogen was the original building block of

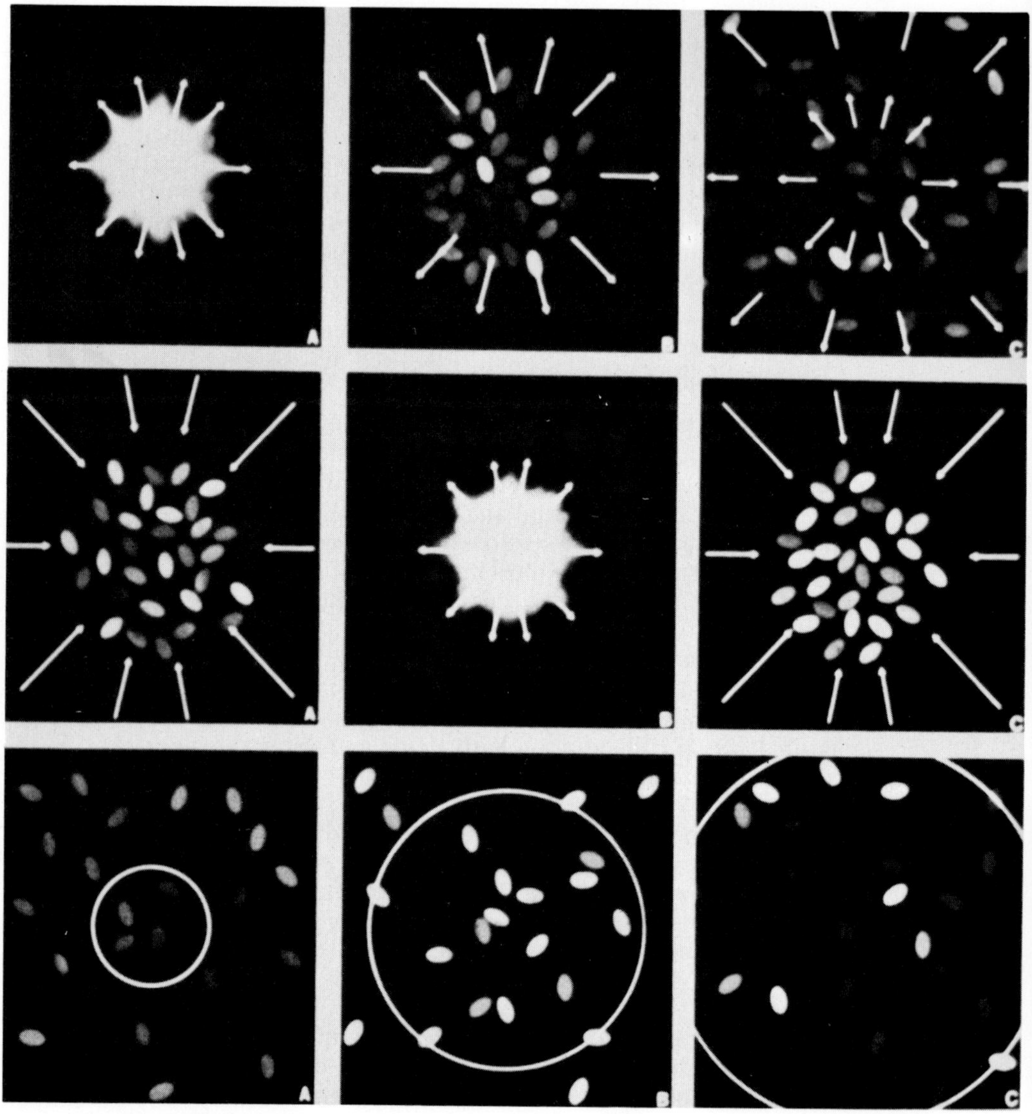

FIG. 54: BIG-BANG THEORY

The big-bang or exploding universe theory (top row) has been evolved by cosmologists to model the creation of the universe. An alternate explosion-implosion (oscillating-universe) theory (center) and a steady-state theory (below) have been offered by other cosmologists as alternative models. (Courtesy: *Atlas of the Universe*, Mitchell Beazley Ltd., 1970).

the universe. All other atoms are formed within dense stars by collision, and shot out into interstellar space by supernova explosions.

The explosion process appears at the top of Fig. 54 in three sketches, A to C, showing the receding elliptical galaxies. It was noted in the previous section that gravitational collapse in a star or galaxy generates an implosion wave. It is also postulated that the universe can implode, giving rise to an exploding-imploding oscillating (pulsating) universe, illustrated in the middle sketches A to C, showing the reciprocating explosion-implosion phases. The cosmic egg then appears momentarily at the point of explosion. An alternative to the two previous imaginative but credulously taxing models discussed above is the steady-state theory of continuous-creation referred to earlier and depicted in the lower sketches A to C. In this equally incredible model, it is postulated that as the galaxies expand, hydrogen is continually being "created out of nothing" to fill the void so that the universe appears the same to an observer for all time and space. The latter "cosmological principle" was advanced by Milne. Consequently, there can be no "moment of creation" and the universe is forever unchangeable. Regardless of the model chosen, unfortunately, they are not nearly as exciting or inspiring as the dramatic words, "let there be....", uttered in Genesis. Undoubtedly cosmologists will do better in the future.

5

PEACEFUL USES OF CHEMICAL AND NUCLEAR EXPLOSIVES

It is beyond the scope of this book to cover the engineering uses of explosives that men have devised over the past century. Numerous papers and texts on the subject have been written, some of which are listed in the references. A number of uses were described previously in the subsection entitled Gunpowder.

Varied Industrial Applications of Chemical Explosives

Aside from the uses described earlier, machinery foundations and transmission-line pole-sinking for the mechanical and electrical engineering industries can be mentioned. An important technology has developed around explosive working, welding, cutting, cladding, embossing and engraving of metals to produce complex shapes and sophisticated parts.

It is of interest to point out that the consumption of industrial explosives in the U.S.A. now is about a million tons (one megaton) per year. Most of the uses are in coal mining (35%), metal mining (21%), quarrying (20%), railway and other construction (21%), and seismographic work (2%). All other sources account for only 1%.

One of the largest chemical explosions was set off to displace 700,000 tons of rock and water when the tops of two underwater peaks were blown off at Ripple Rock, Seymour Narrows (near the mid-point of Vancouver Island), Canada (Fig. 55). The peaks were a serious hazard to navigation; 114 men were lost and more than 125 vessels had been wrecked over the years. The project required the building of three shafts (two about 50 feet below the water level and 300 feet deep, and a service shaft 570 feet deep) all connected by 2150 feet of tunnel, and 1400 tons of explosives costing nearly 3 million dollars to accomplish the job. On 5 April, 1958, the blast sheared off the peaks to a 50-foot depth, making the narrows safe for shipping. At that time, this was the largest non-atomic blast in history. However, it has since been exceeded by some Russian earth-moving projects where up to 3600 tons of explosive were used to make a dam across a gorge near Alma Ata on 21 October, 1966. Teller et al cite another Russian project in which a total of 27,000 tons of explosives were used in 1964; and Toman notes that 186,000 tons were used in 1968 on a single project in the U.S.S.R. to move 2.0×10^6 cubic yards of rockfill while constructing a dam.

A recent industrial innovation has been the application of electro-explosive devices in the space exploration program. For example, 22 of these devices were used on the Gemini program and more on the Apollo program. Such pyrotechnic devices (or energy amplifiers) rely on a low-energy electrical pulse to an exploding wire or heated wire to

FIG. 55: A CHEMICAL BIG BANG
One of the largest non-nuclear blasts was set off at Ripple Rock, Seymour Narrows, B.C., in 1958, where more than 125 vessels were wrecked and 114 men lost before the peaks were blasted away to a depth of 50 feet by 1400 tons of explosives at a cost of nearly $3 million. Some 700,000 tons of rock and water were hurled aside by this explosion. The waters have since been safe for navigation. (Courtesy: du Pont Magazine).

trigger a high-energy release from an explosive charge (or propellant). This energy is then used to pressurize, push, pull, rotate, eject, cut, shear or perform some other function. These devices are reliable, fast-acting and their firing can be timed to a high degree of precision. Their reliability is of paramount importance, since the success or failure of a mission often hinges on their operation. This was dramatically illustrated in the Apollo 13 flight, where explosive primacord was used to cut the lunar excursion module away from the command module, a vital operation in the complex sequence that ultimately brought the astronauts safely back to earth. Members of the Staff from UTIAS, including the author, were consulted prior to this successful operation.

Imaginative Peaceful Potential Uses of Nuclear Explosives

Nuclear explosives differ markedly from chemical explosives, particularly when one considers that the former release radioactive debris into our atmosphere, our waters, and underground, with serious biological damage a possible result. (For example, on 18 December, 1970, about 600 workers had to be evacuated from the Nevada Test Site, when an underground nuclear explosion of less than 20 kilotons spewed a dust plume 8000 feet into the air. About 300 of the men were found to have had their clothing and vehicles radioactively contaminated. Before applications of nuclear explosives are publicly accepted as tolerable, (something that is difficult to evaluate precisely), radiation limits will have to be established in addition to those set for blast and seismic hazards on existing facilities. So far, most of the experience in this area has been obtained from project Plowshare, which was initiated about 1957. In 1959 a series of row charges were fired in New Mexico by the Sandia Corporation to develop cratering techniques for canals, railroads and highways. These tests were extended over the years to cover several types of materials in different locations (clay, coarse alluvium, granite, salt, dolomite, basalt, and tuff, Fig. 56). A wave diagram of the explosion emplaced 636 feet below the surface, and six schematic sequential diagrams of events are sketched. In the time (milliseconds) versus distance R (meters) diagram, the shock wave is seen to race to the surface at rapidly-decaying speeds. When it reaches the surface, the high pressure behind it must be reduced to the value of one atmosphere which exists at the surface. This is brought about by a reflected rarefaction wave, which moves into the ground to relieve the pressures and to interact with the hot cavity that has grown to a radius of about 250 feet. The sketches depict the following: a) detonation of device, temperature about 10 million °C, pressure approximately 1000 million atmospheres, formation of vapour cavity and blast wave (dashed circle) accelerates material outward (point A moves to A'); b) shock wave reaches surface and rarefaction wave reflects, giving rise to spall (fracturing); motion of typical points A to A' and B to B'; solid circle indicates the limit of plastic yield, c) rarefaction wave reaches cavity, which grows upwards and outwards up to 200 milliseconds, d) and e) massive venting up to one second, f) ejecta has returned to

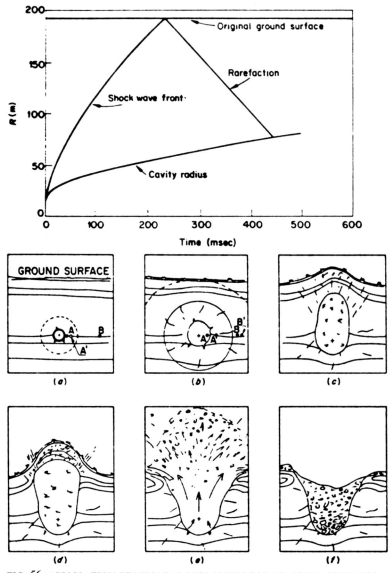

FIG. 56: APPLICATION OF NUCLEAR EXPLOSIVES FOR PEACEFUL PURPOSES
Project Sedan used a 100 kiloton nuclear explosion for the Plowshare program in order to test the concept of trench formation. (Courtesy: *The Constructive Uses of Nuclear Explosives* by E. Teller et al. Copyright 1968 by McGraw-Hill Inc., with permission).

the crater, which is 1200 feet diameter by 230 feet deep, seconds after the detonation.

Cost and compactness are attractive advantages of nuclear explosives over chemical explosives. Teller et al have pointed out that at the 10 kiloton level the cost difference is one order of magnitude (TNT: $460 per ton; nuclear: $35 per ton) whereas at the 2 megaton level, it is three orders less for the nuclear explosion (30 cents per ton). In addition, the 2 megaton TNT charge would require a cube 320 feet per side, and four years of production by the entire U.S.A. explosives industry operating at the present rate, to produce it, whereas only a 40-inch diameter drilled hole would suffice to admit passage of the nuclear explosive. It is worth noting that the costs of nuclear explosives do not vary to any great degree in the 10 kiloton to 12 megaton range. Each would cost about 1/2 million dollars. Only a 40-inch diameter hole would be used for emplacement of devices within this range. However, these bare costs are not, in themselves, indicative of the actual costs incurred in analysis, planning, handling, safety measures, population relocation, and other problems associated with the use of nuclear explosives on large projects. Apparently, it is still cheaper to use large earth-moving equipment rather than chemical explosives to move large amounts of material. Consequently, a real-cost study should compare nuclear explosives with the most efficient conventional methods.

In 1967, a 13-foot device, having a yield of about 26 kilotons, was detonated underground at a depth of 4240 feet about 55 miles east of Farmindale, New Mexico. The project was designated Gasbuggy and was designed to spur the recovery of natural gas. It was the first commercial test in the Plowshare program designed to develop peaceful uses for nuclear explosives. The cost was estimated at $4.7 million. It was quite successful and several other gas-stimulation experiments, with much larger explosions, have subsequently been made in other parts of the U.S.A.

Additional uses have been proposed such as earth moving for canals, harbours, dams, reservoirs, aqueducts, railroads, highways, and quarries; removing the overburden from mineral deposits; producing usable reservoirs in marginal gas and oil deposits in shale that could not otherwise be profitably worked; the on-site (*in-situ*) leaching of impoverished ores; power generation from geothermally heated dry-rock, which is fractured by nuclear explosives to form a cavity for steam generation.

Figure 57 illustrates the formation of a basic chimney rather than a crater, as shown in Fig. 56. The outward motion of the shock wave at 3 and 50 milliseconds is clearly shown. The enlargement of the cavity and the fragmentation of the granite, followed by the final formation of the chimney, is well sketched. Once this basic configuration is formed, it can be used to collect gas from otherwise unprofitable gas-producing areas (Fig. 58); as a means for leaching (washing) ore in order to recover valuable metals (Fig. 59); and as a device for generating steam from the geothermal heat of a cavity to produce electrical power (Fig. 60). Geothermal heat may well become a major energy resource of the *near future* because of depleting gas and oil reserves. As was noted previously, thermonuclear energy may well be the *long-term* answer to all of our energy needs. In this regard, Fig. 61 shows the application of nuclear explosives to oil recovery from shale, while Fig. 62 illustrates a similar undersea scheme as well as a storage concept.

A very important application is that of storing radioactive waste from our fission nuclear reactors. This waste will be a threat to all living things for generations to come unless it can be stored without danger of accidental release due to artificial or natural causes. A scheme for this purpose, using a 5 kiloton nuclear explosive device, is

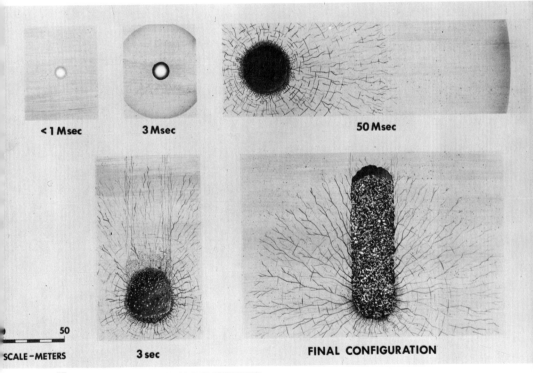

FIG. 57: FORMATION OF A NUCLEAR CHIMNEY

The use of nuclear explosives for peaceful purposes in Project Plowshare for forming basic underground chimneys which can be utilized for many purposes, is illustrated in Figs. 58 to 60. This sketch shows the formation of a chimney, using a 5 kiloton nuclear explosive in granite, as a function of time, from less than a millisecond to several seconds. The final chimney is about 110-feet in diameter and 400 feet long. (Courtesy: U.S. Atomic Energy Commission).

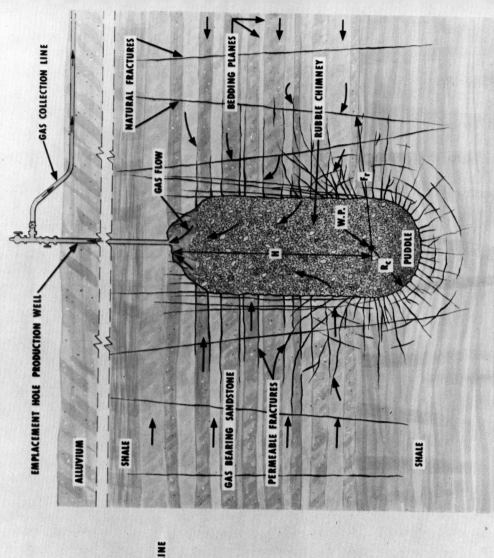

EMPLACEMENT HOLE PRODUCTION WELL

GAS COLLECTION LINE

NATURAL FRACTURES

BEDDING PLANES

RUBBLE CHIMNEY

GAS FLOW

W.P.

F_r

R_c

H

PUDDLE

ALLUVIUM

SHALE

GAS BEARING SANDSTONE

PERMEABLE FRACTURES

SHALE

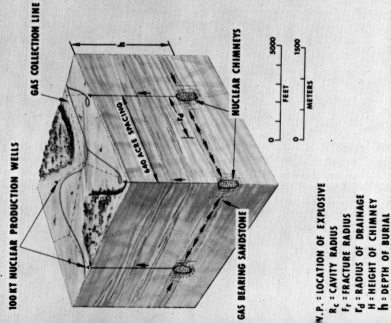

100 KT NUCLEAR PRODUCTION WELLS

GAS COLLECTION LINE

640 ACRE SPACING

r_d

h

NUCLEAR CHIMNEYS

GAS BEARING SANDSTONE

FEET
0   5000

METERS
0   1500

W.P. = LOCATION OF EXPLOSIVE
R_c = CAVITY RADIUS
F_r = FRACTURE RADIUS
r_d = RADIUS OF DRAINAGE
H = HEIGHT OF CHIMNEY
h = DEPTH OF BURIAL

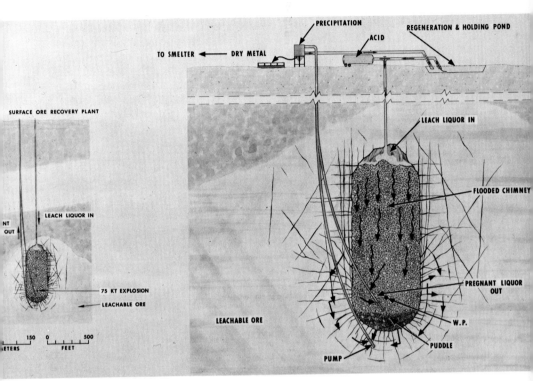

SURFACE ORE RECOVERY PLANT

FIG. 59: THE USE OF NUCLEAR EXPLOSIVES FOR ORE LEACHING
Another important application for underground chimneys formed by nuclear explosives is ore leaching. (Courtesy: U.S. Atomic Energy Commission).

shown in Fig. 63. The nuclear wastes are pumped into the cavity over a period of 25 years. The heat from the radioactive wastes melts the surrounding rock over the next 64 years to a maximum radius and the rock then subsequently solidifies to freeze in the radioactive wastes. This excellent concept will have to await public acceptability before it is applied.

Figure 64 shows how the method of cratering (Fig. 56) can be applied to create artificial harbours, especially on continental shelves where the water is shallow and a rock-covered sea bed makes normal excavation costly.

Although these uses are impressive, and their technological execution is well-understood and apparently cheaper than chemical explosives, factors of safety and public acceptability will first have to be

FIG. 58: NUCLEAR STIMULATION
An important application of the underground chimney formed by nuclear explosives is the recovery of natural gas, from fields with marginal production. This is a particularly important method for the next few decades as fossil fuel reserves dwindle and become in short supply. (Courtesy: U.S. Atomic Energy Commission).

resolved satisfactorily. As a consequence, it does not appear likely that nuclear explosives will be used on a significant scale for these commerical applications in the very near future. In this regard, the question of public acceptability is very similar to the current controversies over issues such as sonic boom, radiation contamination from stationary fission-power reactors, traffic volume and noise, high-rise structures and other environmental problems.

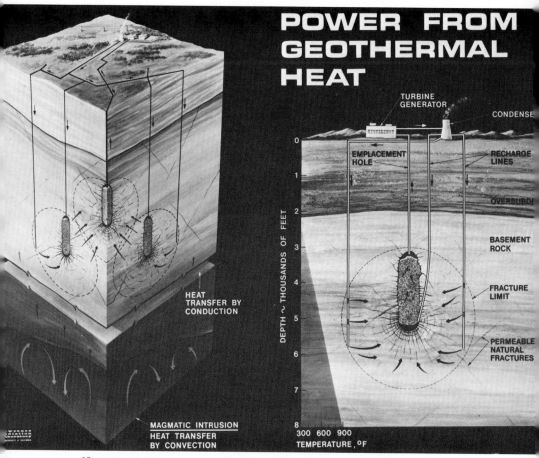

FIG. 60: GEOTHERMAL ENERGY FROM FRAGMENTED DRY HOT ROCKS

A very significant application of an underground chimney formed by nuclear explosives is for geothermal electric-power generation. The heat content of the fragmented dry hot rock is extracted by pumping down water under pressure to form steam, which is then used to drive turbine-generators. Geothermal energy may be a very important resource of the future. (Courtesy: u.s. Atomic Energy Commission).

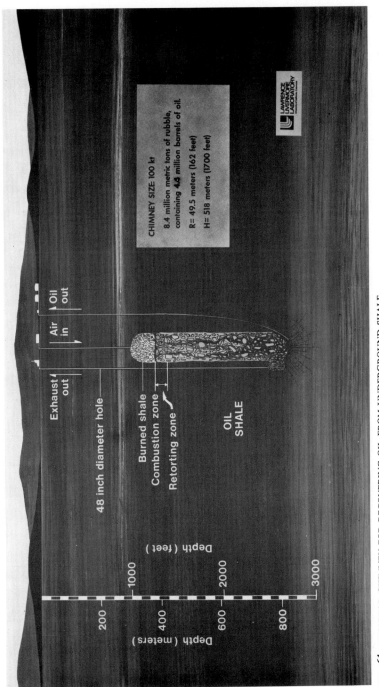

FIG. 61: NUCLEAR CHIMNEY FOR RECOVERING OIL FROM UNDERGROUND SHALE

Great oil reserves exist in huge shale deposits in the U.S.A. and in the Canadian tar sands. In this era of rapid fossil fuel depletion, a scheme is offered for using a nuclear explosive to produce a chimney filled with crushed shale which can be burned from the top down to recover the valuable oil *in situ*. (Courtesy: U.S. Atomic Energy Commission.)

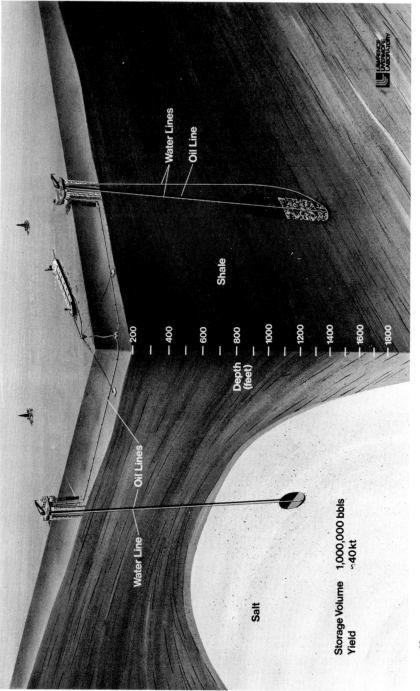

FIG. 62: NUCLEAR CHIMNEY FOR RECOVERING OIL FROM SHALE UNDERWATER AND ITS STORAGE
A variation of the scheme outlined in Fig. 61 is shown here for sea operations over large deposits of shale. A 40-kiloton nuclear explosive could provide storage for one million barrels of oil in a salt-bed cavity. (Courtesy: U.S. Atomic Energy Commission).

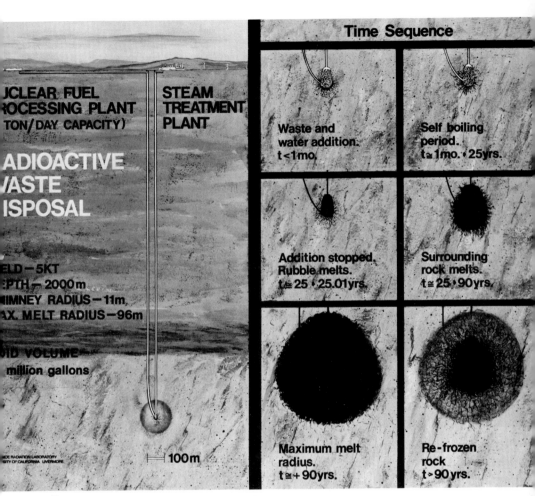

Time Sequence

JCLEAR FUEL
ROCESSING PLANT
(TON/DAY CAPACITY)

STEAM
TREATMENT
PLANT

ADIOACTIVE
VASTE
ISPOSAL

ELD — 5KT
EPTH — 2000m
HIMNEY RADIUS — 11m,
AX. MELT RADIUS — 96m

ID VOLUME
million gallons

NCE RADIATION LABORATORY
SITY OF CALIFORNIA LIVERMORE

100m

Waste and
water addition.
t < 1 mo.

Self boiling
period.
t ≅ 1mo. — 25yrs.

Addition stopped.
Rubble melts.
t ≅ 25 – 25.01 yrs.

Surrounding
rock melts.
t ≅ 25 – 90 yrs.

Maximum melt
radius.
t ≅ + 90 yrs.

Re-frozen
rock
t > 90 yrs.

FIG. 63: SAFE STORAGE FOR NUCLEAR WASTES

The initial problem of safely storing radioactive wastes from nuclear reactor stations has yet to be acceptably solved. An imaginative solution is to use a chimney produced by a nuclear explosive (sketch on the left). The time sequence (six sketches on the right) show how the wastes are admitted over a period of 25 years, additional enlargement of the cavity from the radioactive heat to a maximum size at 90 years and the subsequent freezing of the wastes into the rock for unlimited safe storage. (Courtesy: U.S. Atomic Energy Commission).

FIG. 64: BUILDING A HARBOUR WITH NUCLEAR EXPLOSIVES
The application of the trench-formation technique shown in Fig. 56 to produce a harbour in the continental shelf, where other excavation techniques may prove very difficult, is another good example of the possible future use of nuclear explosives. (Courtesy: U.S. Atomic Energy Commission).

6

PLANAR, CYLINDRICAL AND SPHERICAL EXPLODING AND IMPLODING SHOCK WAVES

An overview is given of some of the properties of planar, cylindrical and spherical shock waves produced in the laboratory by bursting pressurized diaphragms. Analytical and physical characteristics of these waves are discussed in gross terms and at a molecular level. Several of the concepts are then applied to TNT and nuclear explosions with an example illustrating their "scaling laws".

Generation of Shock Waves by Moving Pistons and Bursting Diaphragms

Perhaps the simplest way of seeing how a shock wave is generated can be illustrated by observing the motion of a piston in a cylinder (Fig. 65). If the piston is moved gently in a cylinder at time t_1 a disturbance in the form of a planar sound pulse (Mach wave or characteristic line) is generated. It races ahead of the piston at the speed of sound. This pulse slightly heats and compresses the gas, as shown in the upper pressure-distance (p,x) diagram. If the piston is now pushed again at time t_2, another sound pulse is produced, which further heats and compresses the gas. Because this pulse is moving into gas preheated by the first pulse, it will move faster (sound speed increases with temperature) and will overtake and unite (coalesce) with the first pulse to form a very weak shock wave. This process is shown in the middle sketch of Fig. 65. Ultimately, by continuing to move the piston, a strong shock wave is generated at time t_3, from the compression pulses or waves as shown in the lower sketch. From the (p,x)-diagram, it is seen that a step change in pressure takes place at the shock front and is maintained right up to the piston face. Similarly, there are step changes in all of the physical quantities such as temperature, density and particle velocity. In the latter case, in front of the shock wave, the particle velocity is zero and behind it has a value (which is identical with the piston velocity) corresponding to the strength of the shock wave; that is, the higher the piston velocity, the stronger the shock wave.

Behind the piston, the reverse process takes place as the gas is rarefied and cooled. The pulses diverge and give rise to a rarefaction or expansion wave.

If the piston is accelerated impulsively from rest to a finite velocity, a shock wave is instantly generated and all quantities remain constant from the piston face to the shock front. In practice, such impulsive motion is simulated by breaking a thin diaphragm (cellophane or metal) under pressure in a shock tube (to be discussed), to reduce the inertia effects of a piston to a minimum. This process is illustrated in Fig. 66. The fragmented cellophane diaphragm appears on the left. The turbulent, high-pressure gas is seen to be rushing out through the fragments in a jet-like manner. The process of the overtaking compression waves forming into a shock wave (to the right) is clearly

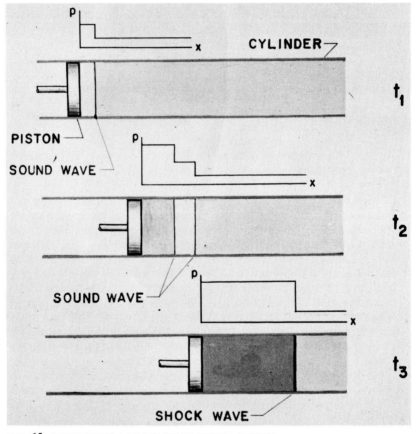

FIG. 65: GENERATING A SHOCK WAVE WITH A PISTON

The production of a planar shock wave from overtaking sound waves (Mach waves) resulting from a gentle piston motion is illustrated by means of the pressure – distance (p–x) plots and the positions of the piston, sound waves and shock wave in the cylinder at three progressively increasing times, t_1 to t_3. (Courtesy: UTIAS).

illustrated. As the high-pressure gas bulges the diaphragm before it bursts, the shock wave takes on a curved shape and it reflects at the solid boundary. This reflection process makes it possible for the shock wave subsequently to become planar.

Although an originally plane, pressurized diaphragm can be used to generate a planar shock wave in a shock tube (a name given to a tube where a diaphragm separates a high-pressure gas in a driver section from a low-pressure gas in a channel section for the purpose of producing shock waves), it is also possible to produce spherical and cylindrical shock waves by using pressurized glass spheres and cylinders as diaphragms. Figure 67 shows such glass diaphragms, while Fig. 68 sequentially illustrates the breaking process of a glass sphere over a time period lasting 850 microseconds. At 200 microseconds, the compressed gas rushes out through the fragments (similar to the

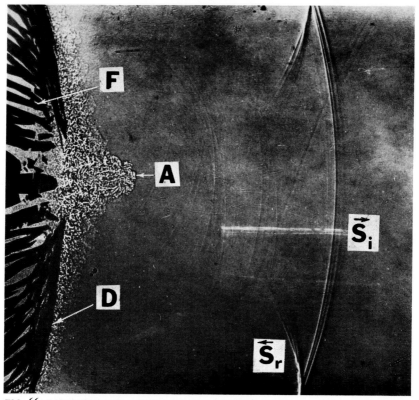

FIG. 66: FORMATION OF A SHOCK WAVE IN A SHOCK TUBE

A shadowgram of the bursting of a "red-zip" cellophane diaphragm in the UTIAS 2×7 inch shock tube at an initial pressure ratio of 2 across the diaphragm. The bulging, ruptured diaphragm D, and the pressurized air A, escaping through the fragments F, appear on the left. The curved incident S_i, and reflecting S_r, compression and shock waves are seen on the right. (Courtesy: UTIAS).

planar case, Figure 66) to form compression waves. At 255 micro-seconds, a shock wave is clearly evident, while at 375 micro-seconds, (identical to Fig. 5) the shock wave is fully formed. At 850 microseconds, the main shock wave has disappeared from view and only the heavier, expanding glass fragments remain (see also Fig. 6 and, by analogy, Fig. 54).

Laboratory Controlled Explosions and Implosions

For the more analytically inclined reader, it may be noted that the wave formation processes for planar, cylindrical and spherical explosions and implosions can best be illustrated on a radius-time (r,t)-diagram shown on Fig. 69. The diaphragm is designated by D, the initial gas states by p_4, and p_1, the main shock wave by S, the contact surface by C, the rarefaction wave by R, the initial radius (or distance to the line of symmetry) by r_0 and the sound speed by a_4. For explosions (diverging shock waves) p_4 is greater than p_1 ($p_4 > p_1$) and the reverse is true for implosions (converging shock waves). The sketches are drawn for small times ($t \ll r_0/a_4$), that is, less than the time taken for the head of the rarefaction wave to reach the t-axis

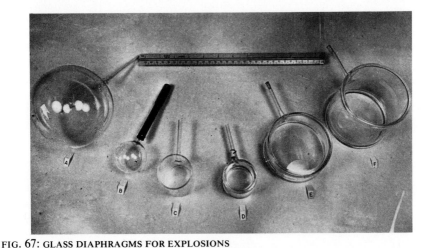

FIG. 67: GLASS DIAPHRAGMS FOR EXPLOSIONS
Some glass spheres and cylinders used for explosion and implosion experiments; (A) 4.9-inch-diameter sphere, wall-thickness 0.04 inches, weighing 2.0 ounces; (B) 1.9-inch-diameter sphere, wall-thickness 0.04 inches, weighing 0.7 ounces, cemented in a holder; (C) 1.9-inch-diameter cylinder 2.6 inches long by 0.06 inches thick, weighing 1.3 ounces with open ends; (D) 1.9-inch-diameter cylinder 1.9 inches long by 0.06 inches thick weighing 1.7 ounces, with 0.125-inch-thick optical flat, closed ends; (E) 3.9-inch-diameter cylinder 1.9 inches long, by 0.125 inches thick weighing 6.4 ounces, closed ends; (F) 3.9-inch-diameter cylinder 3.9 inches long by 0.125 inches thick weighing 9.6 ounces, closed ends. (Courtesy: UTIAS).

(r = 0). This is done purposely to avoid a more complex diagram detailing wave reflections and interactions.

It is seen that for all explosions when the diaphragm is ruptured, a shock wave moves to the right followed by a contact surface, and a rarefaction (expansion) wave moves to the left. The shock wave compresses and heats the low-pressure gas and the rarefaction wave cools and expands the high-pressure driver gas once held back by the diaphragm. The new boundary of the driver gas is now the contact

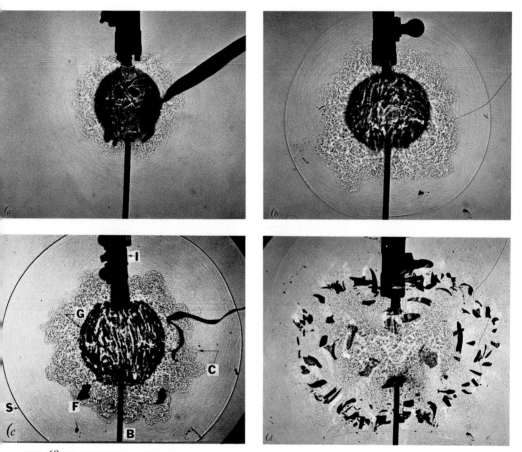

FIG. 68: BLAST FROM A PRESSURIZED GLASS SPHERE
Shadowgrams showing the formation of a spherical shock wave in air following the rupture of a 2-inch-diameter glass sphere pressurized with air to 300 psi (20 atmospheres). In (a) at 200 microseconds, the picture resembles the planar case shown in Fig. 66, where the high-pressure turbulent gas is rushing past the glass fragments to drive compression waves that coalesce to form the beautifully symmetric shock shown in (b) and (c), at 255 and 375 microseconds, respectively. In (d), at 850 microseconds, only the flying glass fragments remain. (Courtesy: UTIAS).

surface or essentially the driving piston of Fig. 65. At this surface, the flow pressure and velocity on either side are matched but all other thermodynamic quantities are different. The expansion and cooling through the rarefaction is a constant entropy (isentropic, reversible) process, whereas the compression and heating process through the shock transition is irreversible and the entropy increases. On either side of the shock front, however, the entropy is a constant. The change in entropy at the contact front is usually manyfold greater than that across the shock wave, in a shock-tube flow. For example, for the case of air in the shock-tube chamber and channel, at a pressure ratio of twenty across the diaphragm (at a shock wave Mach number $M = 1.8$), the change in entropy across the contact surface is thirteen-fold greater than across the shock wave. An increase in entropy is an

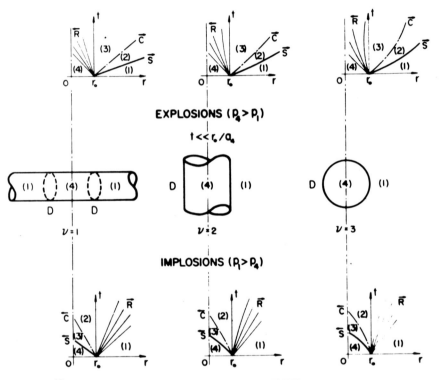

FIG. 69: PROPERTIES OF EXPLOSIONS AND IMPLOSIONS

Schematic diagrams showing time-distance (t–r) wave diagrams for explosions and implosions following the rupture of planar ($\nu = 1$), cylindrical ($\nu = 2$) and spherical ($\nu = 3$) diaphragms. In the case of an explosion, the diaphragm is pressurized internally compared to the surrounding atmosphere ($p_4 > p_1$), for an implosion, the reverse is true ($p_1 > p_4$). The basic wave paths for shock waves (S), rarefaction waves (where each line represents a sound wave, Mach wave or characteristic line) (R) and contact surfaces (C) are clearly drawn. (Courtesy: UTIAS).

indication of energy bound up to maintain the gas at its existing state and is therefore not available to do useful work (see Preface).

In Fig. 69, for the planar case, state (2) lying between the shock wave and the contact surface is well suited for aerodynamic and physical research and testing, whereas state (3) which has been disturbed by the jagged remains of the diaphragm is not useful. For the cylindrical and spherical explosions, states (2) and (3) are no longer uniform and cannot be used for aerodynamic testing purposes.

Each trace on the time-distance diagram represents a wave velocity (those more parallel to the r-axis have very high velocities, that is, a large distance is covered in a small time or the rate of change of r with respect to t, dr/dt is large; those near the t-axis have very low velocities). It is seen that, for the planar wave system, all wave velocities are uniform, whereas for the cylindrical and spherical explosions, the shock waves and contact surfaces, respectively, decelerate more noticeably, and the wave pulses of the rarefaction wave (sound waves, Mach waves, characteristic lines) become curved towards the t-axis. For an implosion, the reverse occurs for cylindrical and spherical flows. The planar case remains unchanged, as in this case we cannot differentiate between an explosion and an implosion (Fig. 69).

This time-distance process can be photographed by using a rotating-drum camera. The wave systems for the planar, cylindrical and spherical cases are shown in the schlieren photographs in Figs. 70, 71 and 72, respectively. The beautiful series of photographs in Fig. 73, complement the time-distance record of Fig. 72. It makes it possible for the reader to compare the usual pictures and clarify what he sees in a continuous-time photograph at identical instants.

The development of the spherical explosion with time from frame 1 to 5 is portrayed in an excellent manner. The second shock wave is seen to emerge at the top of frame 7, and in frame 8 only the driver gas and some glass fragments are visible. The central frame is an overlay of the events; the original sphere and the glass fragments are discernible. In all cases, the diaphragm is a physical obstacle that disturbs the flow. Consequently, the wave system at the origin departs from the ideal sketches of Fig. 69. Nevertheless, as one moves away from this disturbed region, the system approaches the idealized model. In the planar or shock-tube case, however, the walls of the tube also disturb the flow through viscosity and heat conduction. They cause a boundary layer to grow on the tube walls. As a result the shock wave decays and the contact surface accelerates. The boundary layer also disturbs state (2) and the cold region (3), shown in Fig. 69. In fact, at low channel densities where these effects are magnified, the shock wave slows down and the contact front speeds up towards a common terminal velocity and the flow region (2) and (3) in Fig. 69 become nonuniform. Consequently, the useful test times for very high velocities and low densities are limited to microseconds. Despite this,

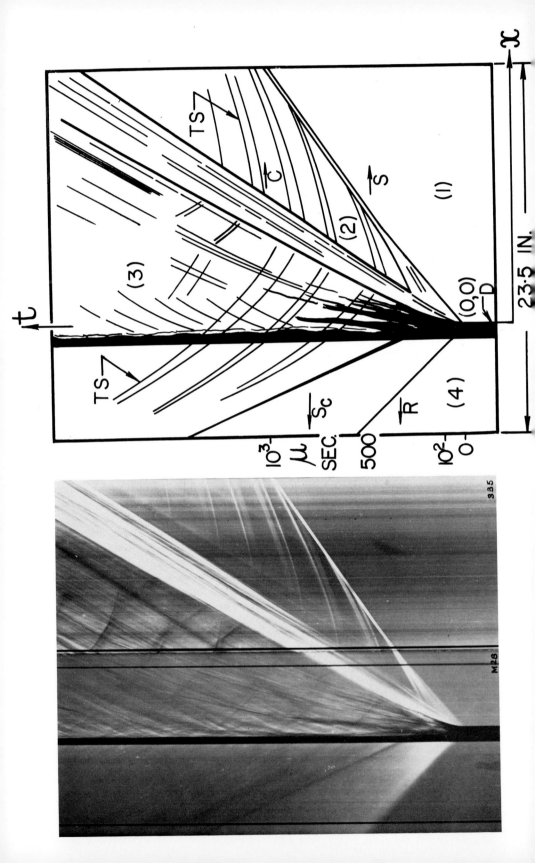

the shock tube and its various derivatives (shock tunnels, expansion tubes) have become the "test-tubes" of modern gasdynamics. It is doubtful if the design of the vital heat shields for the manned-spacecraft programs would have been possible without the use of these devices as aerodynamic test facilities. Figure 74 shows the data points obtained in shock tubes that made such a design possible and protected the life of the astronauts during re-entry. Plotted are the heat transfer rates as a function of flight velocity and re-entry altitude. The agreement of actual shock-wave test data with theory is excellent.

It is worth noting that, for cylindrical and spherical flows, an implosion wave (second shock) is formed, as shown clearly in Fig. 72, which reflects from the origin, goes out as an explosion wave S_2, and follows the main blast wave S_1. Its role in this time-varying flow is similar to the tail shock from a supersonic aircraft (Fig. 39). As a matter of fact, mathematically, the flow generated by a strong cylindrical explosion is analogous to that produced by a slender body travelling at very high speed (hypersonic flight). This concept is illustrated in Fig. 75, where it is seen that as the hypersonic vehicle passes progressive planes in its flight (right to left) it leaves a trace on a plane similar to a cylindrical explosion where the shock wave is driven by the expanding gases or piston. Such concepts are very useful in understanding both types of flows. Weak spherical explosions also have practical applications in building travelling-wave sonic-boom simulators in the laboratory (see Chapter 8).

FIG. 70: PHOTOGRAPH OF A PLANAR EXPLOSION

A schlieren record and explanatory sketch of the wave system in the time–distance (t–x) plane produced in a shock tube from the instant the diaphragm ruptures, shows: R, rarefaction wave, S, shock wave, C, contact front, S_c, condensation shock wave, TS, transverse shock waves, D, bulging cellophane diaphragm; (1), air in low-pressure channel, (4), air in high-pressure chamber; pressure $p_4 = 2.2$ atmospheres, $p_1 = 0.44$ atmospheres, $p_4/p_1 = 5.00$, sound speeds in chamber and channel, a = 1132 feet per second, shock wave Mach number, M = 1.39, pressure rise ratio across shock wave $p_2/p_1 = 2.1$, flow Mach number of region (2) heated and compressed by shock wave $M_2 = 0.50$, flow Mach number in region (3) cooled and expanded by the rarefaction wave $M_3 = 0.65$. (Courtesy: UTIAS).

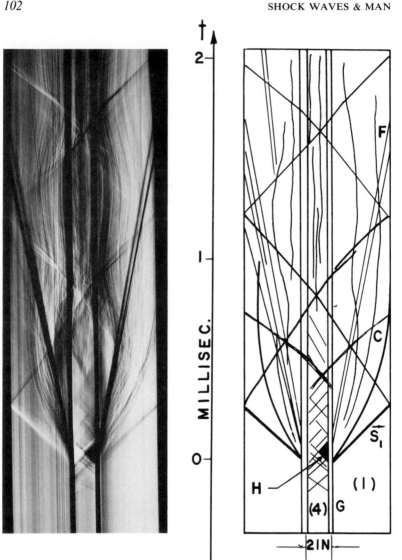

FIG. 71: PHOTOGRAPH OF A CYLINDRICAL EXPLOSION

A schlieren record and explanatory sketch of the wave system produced in the time–distance (t–r) plane, by rupturing a pressurized glass cylinder 2-inches in diameter and 2.5 inches long, shows: G, intact glass cylinder to be ruptured mechanically; F, glass fragments; H, head of rarefaction wave; S_1, main shock wave; C, contact surface; air pressure $p_4 = 24$ atmospheres, air pressure $p_1 = 1.0$ atmosphere, temperatures $T_1 = T_4 = 27°C$. (Courtesy: UTIAS).

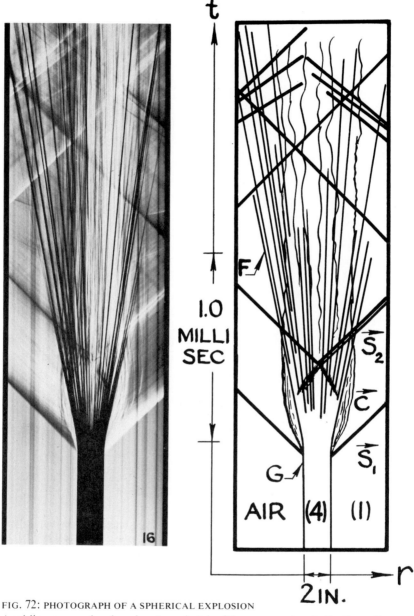

FIG. 72: PHOTOGRAPH OF A SPHERICAL EXPLOSION

A schlieren record and explanatory sketch of the wave system in the time–distance (t–r) plane produced by rupturing a 2-inch-diameter pressurized glass sphere shows: G, glass sphere still intact; F, shattered sphere fragments; S, main shock wave; S_2, second shock wave resulting from the reflected implosion wave; air pressure $p_4 = 22$ atmospheres, air pressure $p_1 = 1.0$ atmosphere, temperatures $T_1 = T_4 = 26°C$. (Courtesy: UTIAS).

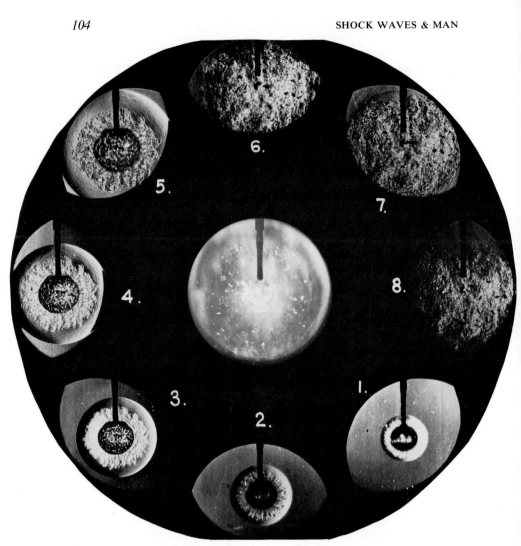

FIG. 73: SNAPSHOTS OF A SPHERICAL EXPLOSION

Eight schlieren photographs of a spherical explosion taken at intervals of 50, 100, 150, 200, 250, 800, 900, 1000 microseconds from frames 1 to 8, respectively, after the 2-inch-diameter glass sphere ruptured. The sphere was pressurized to 38 atmospheres. The blast wave propagated into air at room conditions. The central frame shows the disintegration of the sphere. The pictures complement the less-familiar photography shown in Fig. 72. (See also Fig. 6). (Courtesy: UTIAS).

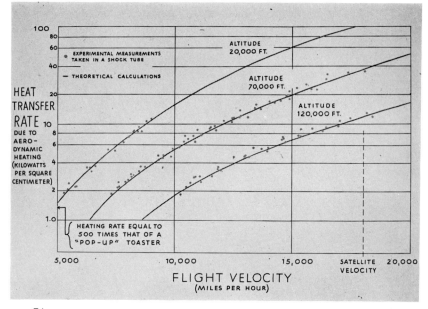

FIG. 74: HEAT-TRANSFER DATA FOR HEAT-SHIELD DESIGN

The heat-transfer rates on Earth re-entry with altitude and flight velocity were obtained in shock tubes. Such data made it possible to design heat shields for the manned spacecraft Mercury, Gemini and Apollo, as well as the Skylab and Space Shuttle programs. (Courtesy: Avco Everett Research Laboratory, Inc.).

Underwater Blast

The technique of using glass spheres as diaphragms can be readily extended to underwater explosions as shown in Fig. 76. The centre frame shows a 3-foot-diameter steel sphere filled with water almost to the top. A glass sphere filled with gas (carbon-dioxide in this case) at high pressure can be seen through the window. As the sphere is ruptured, eight schlieren photographs record the resulting flow patterns at intervals from 0.025 to 24.0 milliseconds. At 0.025 milliseconds, the spherical blast wave in the water is very clear. It then races away at 5000 feet per second to disappear from view. During the interval from 0.50 to 10.0 milliseconds, the lighter gas pushes the heavier glass fragments and the surrounding water slowly but surely away so that the pressure of the gas bubble at a later stage is actually less than that of the surrounding water, since it has been accelerated and is difficult to bring to a stop (the so-called afterflow). Because of this reduced bubble pressure, the motion is reversed and the bubble implodes, leaving the glass fragments behind. If this process were

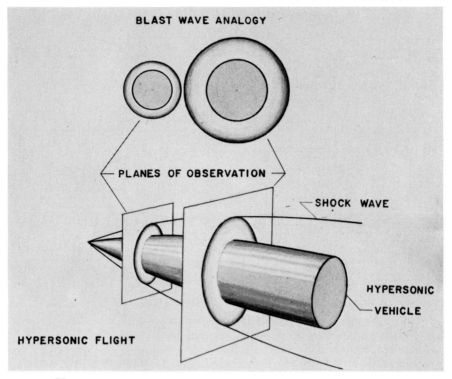

FIG. 75: ANALOGY BETWEEN EXPLOSIONS AND HYPERSONIC FLIGHT

A supersonic or hypersonic aircraft deposits a great deal of energy into the atmosphere by heating and compressing it through the bow and tail shock waves. An observer, viewing the passage of a hypersonic vehicle through a plane, would see it as similar to the motion of a cylindrical blast wave. The two problems can be formulated and solved mathematically using identical methods. (Courtesy: UTIAS).

perfectly reversible, then the shape at 24 milliseconds would be identical to that at 0.025 milliseconds. However, dissipation and wave reflections deform the gas bubble to a kidney shape. The process is then repeated until buoyant and dissipative forces destroy the bubble. The oscillations are reminiscent of one of the models discussed previously (Fig. 54).

Shock-wave Collision

Glass spheres can be used to study more complex interactions such as the collision of spherical shock waves shown in Fig. 77. The central frame shows two spheres hit mechanically by two small metal mallets on a crossbar. The resulting shock waves generated by the two exploding spheres 250 to 600 microseconds (at 50 microsecond intervals) after rupture appear in the eight respective schlieren photo-

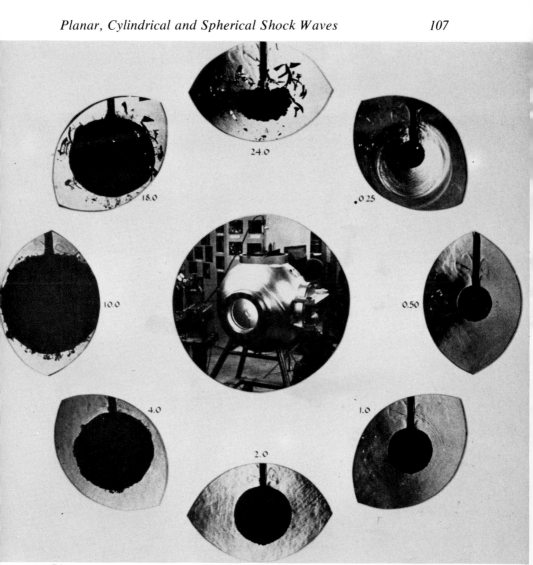

FIG. 76: UNDERWATER BLAST

A spherical underwater explosion is shown in some detail through schlieren photographs obtained from a high-speed multiple-spark camera with a variable framing rate of from 250 microseconds to 24 milliseconds with an exposure time of 2 microseconds. The explosion was generated by bursting a 2-inch-diameter glass sphere filled with carbon-dioxide at 20 atmospheres at a depth of 13 inches in a 3-foot-diameter steel sphere. The water was at 1 atmosphere and 20°C. The times for each frame are given in milliseconds. The central frame shows the intact glass sphere in the water-filled steel sphere prior to the blast. The first frame (upper right) shows the spherical shock wave in water. The remaining frames show the growth and collapse of the oscillating gas bubble as well as the glass fragments. (Courtesy: UTIAS).

graphs. It is seen that the shock waves approach, collide, interpenetrate and recede, leaving only the original expanding gas and the glass fragments visible in the last frame. This type of interaction may occur on a cosmic scale in the collision of nebulae.

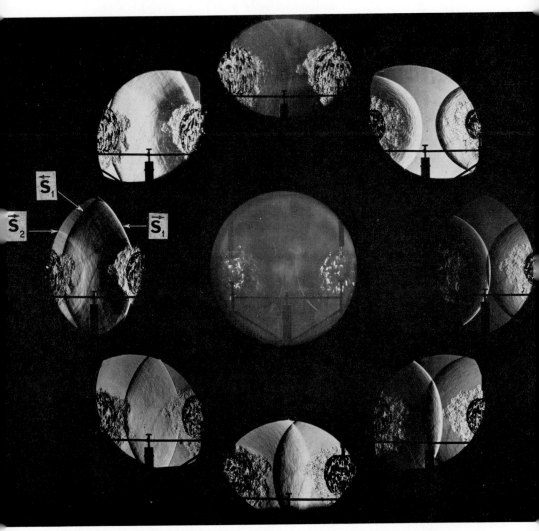

FIG. 77: SNAPSHOTS SHOWING COLLISION OF TWO SPHERICAL SHOCK WAVES

Schlieren records from a high-speed multiple-spark camera show in the central frame two ruptured 2-inch-diameter glass spheres with their centres 9 inches apart initially filled with helium at 22 atmospheres and 22°C. The approach of the two shock waves shown in the upper right frame was taken 250 microseconds after impact by the breaker. The remaining frames, at 50-microsecond intervals, record the collision and penetration of the first S_1, and second S_2, shock waves. The central frame also shows the breaker mechanism. (Courtesy: UTIAS).

The pictures shown in Fig. 77 are complemented by the time-distance (t-r) schlieren photograph of Fig. 78. In this case the right sphere burst 160 microseconds later giving an asymmetic collision with the left shock wave, which has decayed during this period. It is therefore bent more than the right shock wave at the point of interaction. The compressed two-inch helium spheres expand to about 27 times their original volumes and then remain quite stationary except when hit and recompressed by the receding shock waves.

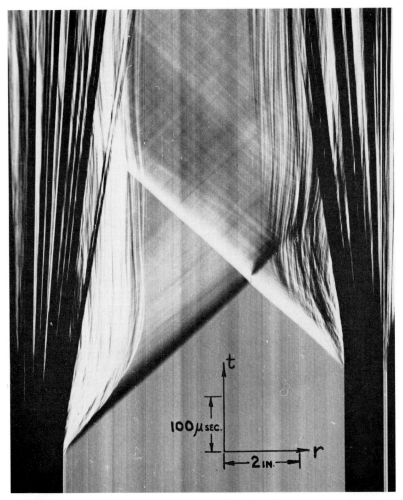

FIG. 78: COLLISION OF TWO SPHERICAL SHOCK WAVES

A continuous schlieren photograph complementing Fig. 77, with the same initial conditions. The time and distance scales appear on the photograph. Note that the sphere on the right ruptured 160 microseconds later than the left sphere and gave rise to a collision of two spherical shock waves of unequal strength (explanatory sketch in Fig. 72 is also applicable here). (Courtesy: UTIAS).

The production of explosions and implosions can be dramatically illustrated by using a sheet of explosive, as shown in Fig. 79. Eight detonators are placed symmetrically about a 10-inch diameter by 0.165-inch-thick sheet of explosive (du Pont EL-506). When these detonators are fired simultaneously, eight circular detonation (shock) waves (see Fig. 104) are generated that race ahead and collide (frames 1, upper left, to 15, lower right). The shocks are strong and, on collision, the explosive gas undergoes additional heating and compression by the interpenetrating shocks, consequently the gas glows and continues to do so over a larger area forming an eight-pointed "star". In the meantime, the outer elements of the shock wave diverge and the inner portions converge as an implosion. At the points of convergence and their reflections (frame 10), unlimited pressures and temperatures are achieved ideally (in practice conduction, radiation and viscous effects limit them to very high but finite temperatures and pressures). As a result, the focus of the implosion glows even more brightly. However, as the colliding, imploding-wave segments are 22.5 degrees out of phase with the original collisions, a new eight-pointed star is formed, which soon outshines the old one (frame 15).

Molecular Structure of Shock Waves in Gases

It will be helpful at this point to examine the shock wave itself in order to understand some of its properties. Figure 80 shows four schlieren photographs of a plane shock wave (S_i, white line) moving to the left (upper right) and colliding with a block. After collision, part of the shock wave (upper left) keeps on going while the remainder is reflected and moves to the right (S_r, black line). The reflection process disturbs and curves the "legs" of the shocks. Similarly, when a planar shock wave S, diffracts over a trench, the "leg" becomes curved as it negotiates the first corner. This illustrates some important aspects of (nonlinear) shock-wave reflection and diffraction, which are unlike linear acoustic phenomena. Perhaps the simplest example of linear reflection in acoustics or optics is that the ray angles of incidence and reflection are identical. In nonlinear shock-wave reflection this is no longer true. The angle of reflection is different from the angle of incidence.

The shock waves themselves, although they appear photographically to have finite widths, are really very thin (about one millionth of an inch at room conditions), but are spread by optical refraction to have the visibly thicker appearance. The ambient molecules, when they are hit by a shock front, undergo collisions to achieve a new equilibrium state behind the front, with increases in pressure, temperature, density and entropy (thermodynamic quantities), as well as an induced particle velocity (wind) behind a moving shock wave.

For the more analytical reader it may be noted that the diatomic molecule (in this case hydrogen) is depicted as consisting of two

FIG. 79: COLLISION OF DETONATION WAVES

A beautiful illustration of generating explosion and implosion waves was obtained by simultaneously triggering eight detonators on the rear surface of a disc of explosive (du Pont EL-506) 10-inch-diameter by 0.165 inches thick. The framing camera took the sequence of 15 photographs (upper left to lower right) at a rate of 600,000 frames per second with an exposure time of 0.6 microseconds. (Courtesy: U.S. Naval Weapons Laboratory).

atoms joined by a stiff spring. At lower temperatures, this dumbell can translate (t) along 3 directions (or degrees of freedom) (x, y, z) and rotate (r) about 2 axes (y and z). The rotational energy about the x-axis is minute, since the moment of inertia in this direction is

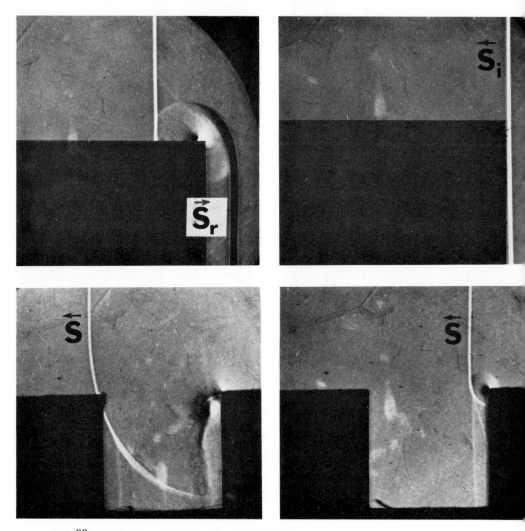

FIG. 80: APPARENT THICKNESS OF SHOCK WAVES

The schlieren photograph (upper right) of an incident shock wave S_i (moving left) just colliding with a block in a 2×7-inch shock tube is an excellent example of apparent shock-wave thickness resulting from optical refraction. The actual thickness is only a few millionths of an inch. The shock diffraction and reflection (S_r) process is also illustrated (upper left). A diffraction process over a trench appears in the lower two photographs. (Courtesy: UTIAS).

negligible. At high temperatures, the spring becomes soft and the hydrogen molecule can vibrate (v) storing kinetic and potential energy (2 degrees). The spring may even break (dissociate, d). At still higher temperatures, an electron, through collision with another electron or atom, will be pushed up from its rest state orbit to a new orbit (electronic excitation) and if the collision is energetic enough, the electron is torn away from the atom (ionizes, i), which now becomes a positive ion. The energies stored can be summed (see Fig. 81), where R is the gas constant per unit mass of gas, T is the absolute temperature (each degree contributes $1/2\,RT$), α is the percent dissociation, l_d the energy of dissociation per unit mass, $\Sigma N_i\epsilon_i$ is the summation of the electron energies in each orbit, x is the percent ionization, and l_i is the ionization energy per unit mass. It should be noted that an excited gas radiates (at a frequency ν) as it falls from a higher to lower orbit (see diagram) separated by an energy ΔE, such that the relation $\Delta E = h\nu$ is satisfied, where h is the Planck constant. That is, as an electron of atomic hydrogen falls from its first excited state to its rest state (orbit) it radiates a red line at 6563 Ångstroms, the so-called hydrogen alpha line (H_α). Consequently, for the case of weak shock waves in air, the molecular collisions (see subsection on re-entry) through the

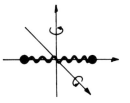

TRANSLATION
AND
ROTATION

VIBRATION

$$H_2 \rightleftharpoons 2H + l_d$$

DISSOCIATION

$\Delta E = h\nu$
For H_α $\nu = 4\cdot57 \times 10^{14} (\lambda = 6562\cdot8\,\text{Å})$

ELECTRONIC EXCITATION

$$H \rightleftharpoons H^+ + e + l_i$$

IONIZATION

$$\text{TOTAL} \quad e = e_t + e_r + e_v + e_d + e_e + e_i$$
$$= \tfrac{3}{2}RT + \tfrac{2}{2}RT + \tfrac{2}{2}RT + \alpha l_d + \Sigma N_i\epsilon_i + 2x\,l_i$$

FIG. 81: ENERGY STORAGE IN A DIATOMIC MOLECULE
The hydrogen molecule is used as a model to show how the shock wave converts the directed molecular energy of motion into thermal and chemical energies which are then stored in translation, rotation, vibration, electronic excitation, dissociation and ionization. (Courtesy: UTIAS).

shock wave divide the thermal energy among the translational and rotational modes ($e_t + e_r$) and the gas is called perfect. However, as shown in Fig. 81, the nitrogen and oxygen molecules at high temperatures (strong shock waves) also have the following additional means of absorbing energy: vibration (e_v), dissociation (e_d), electronic excitation (e_e) and ionization (e_i). Therefore, when an ambient gas like air is hit by a strong shock wave around a re-entering spacecraft, the translational and rotational modes achieve equilibrium in two or three molecular collisions, vibration in hundreds of collisions and dissociation, electronic excitation and ionization in many thousands. The large initial energy which was originally imparted to the translational and rotational motion is now shared, as indicated by the law of equipartition of energy, (1/2 RT) with the other modes of excitation described above. As this adjustment process requires many collisions, the shock thickness is increased (see Fig. 82). During this stage, when the thermal energy is distributed (relaxation time), the temperature drops dramatically. Consequently, the problem of re-entry is greatly alleviated, (see Fig. 45) so that only about one percent of the heat goes into the heat shield and the rest is dissipated through the bow and tail shock waves into the wake of the spacecraft. Were this not the case, the problem of re-entry would be formidable.

Such a relaxation process in a shock front is beautifully illustrated in Fig. 82, which is an interferogram (a special type of photographic plate obtained from an interferometer which is sensitive to the changes in density or refractive index of a gas) of a shock wave in argon, a monatomic gas, moving to the left. It is evident that there is practically a step-change of each fringe as the translational mode is excited. Additional atomic collisions tear off electrons and the fringes gradually shift downward (since electrons have a negative refractive index). Then, when a lot of electrons (avalanche) are torn off by the more efficient electron-electron impact, the fringes suddenly shift down and then straighten (after some oscillations from other causes). The relaxation length produced on approach to equilibrium is about 1 inch, many thousand-fold thicker than the initial translational front.

The foregoing is illustrated graphically for shock waves in gases by the sketches shown in Fig. 83. The relaxation zone due to translation and rotation is shown as L_1 and that due to all other excitations as L_2. It is seen that the sharp rise to the perfect-gas (or the so-called Rankine-Hugoniot) values through L_1 (it is actually in the form of a shallow S-curve rather than a linear rise as sketched) is followed by an exponential approach to the imperfect gas values by an increase in pressure, p (modest) and density, ρ (sizeable) and an exponential decrease in temperature, T (large) and velocity, V (large). It should be noted that this velocity V, is the so-called steady-flow value encountered in supersonic flight or in a wind tunnel where the shock wave position is fixed. In this case, the flow velocity in front of the planar shock wave, V_1, is always supersonic and that behind it, V_2, is always subsonic. In the case of a moving normal shock wave, which always races into the gas at rest at supersonic speeds, it induces a particle velocity behind it that can be subsonic (flow Mach number less than unity, $M < 1$), transonic (nearly unity, $M \simeq 1$) and low supersonic (greater than unity, $M > 1$), with increasing shock strength. The particle velocity is always less than the shock-wave velocity for all

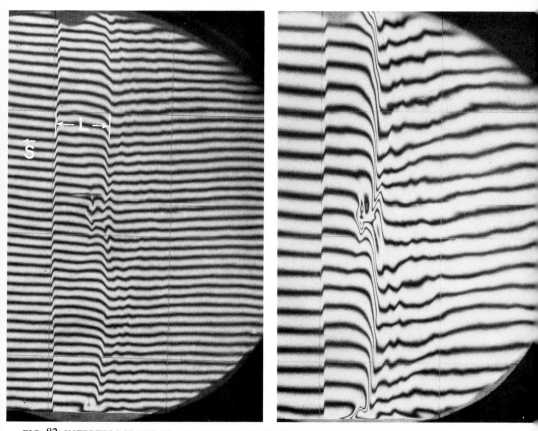

FIG. 82: INTERFEROGRAMS OF A PLANAR SHOCK WAVE

Two laser interferograms, taken simultaneously at wavelengths λ of 3471 Ångstroms (left) and 6943 Ångstroms (right), of a shock wave S, moving left at a Mach number $M = 17.5$ into argon at an initial pressure $p = 0.003$ atmospheres. A sudden rise is seen in the fringes corresponding to the shock fronts, where the 3 translational degrees of the atom are excited. The fringes drop gradually due to the formation of electrons by atomic collisions and suddenly due to an avalanche of electrons from electron-electron impact (very prominent for the longer wavelength, right, which is more sensitive to the electron refractive index). From these two photographs it is possible to measure the electron number density and the total density of the plasma as well as the ionization relaxation length L, or time. (Courtesy: UTIAS).

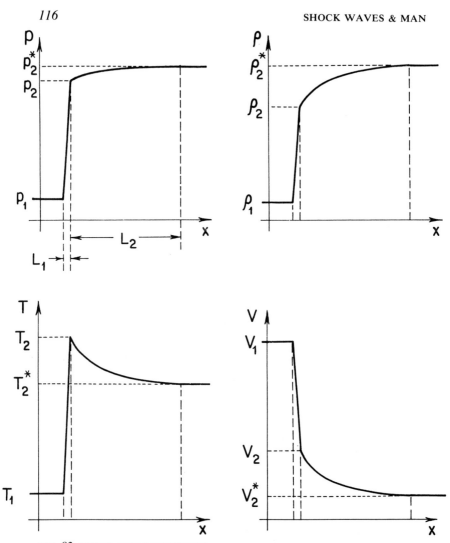

FIG. 83: SHOCK-WAVE THICKNESS

A representation of the transition properties through a shock-wave front. The transition involving the translational and rotational degrees (L_1) is thin (about one microinch at atmospheric conditions). It is also called the Rankine-Hugonoit jump transition for a perfect gas (named after two great fluid dynamicists). It only requires a few molecular collisions to bring the old state (1) to the new state (2). Large increases in pressure (p_1 to p_2), density (ρ_1 to ρ_2) and temperature (T_1 to T_2) and a drop in steady flow velocity (V_1 to V_2) occur. The other modes such as vibration, etc. require hundreds or thousands of collisions. This increases the shock thickness manyfold by L_2. In this distance the gas relaxes to its final equilibrium imperfect gas state in an exponential manner: causing a small increase in pressure (p_2 to p_2^*), a very large drop in temperature (T_2 to T_2^*), a large rise in density (ρ_2 to ρ_2^*) and a significant drop in velocity (V_2 to V_2^*). (Courtesy: UTIAS).

real gases, but it approaches that of the shock for gases with specific heat ratios nearing unity for very strong (high Mach number) shock waves. The moving shock wave can be readily turned into a stationary shock wave by superimposing a counter velocity on the shock wave that just keeps it fixed. The mathematical solutions for the flow quantities then become identical.

Properties of Normal Shock Waves

With these illustrations of the difference between perfect and imperfect gases produced by shock waves with increasing strength, it is possible to consider the properties of planar (or normal) shock waves in a perfect and imperfect gas as shown in Fig. 84. The variations in pressure, density and temperature ratios (p_2/p_1, ρ_2/ρ_1, T_2/T_1) across the shock front are shown as functions of shock Mach number, from M = 1 (where all ratios are unity) to about M = 24 (a re-entry from a circular orbit at a velocity of about 26,000 feet per second) and various initial pressures [0.1 to 760 millimeters of mercury [mm Hg] or one atmosphere]. The line marked $\gamma = 1.4$ (where γ is the ratio of the specific heats at constant pressure to constant volume) considers air as a perfect gas (translational and rotational modes alone are excited). The other lines have all the other modes excited to a degree that increases with rising shock Mach number and decreasing initial pressures for an imperfect gas. The graph shows that the pressure increases only slightly for the imperfect gas, but the density increases greatly and the temperature decreases dramatically from those of a perfect gas. (The density ratio of an idealized perfect gas like air approaches a limit of six, but temperatures and pressures can rise without limit for very strong shock waves.) Similar results exist for reflected shock waves in shock tubes and for the oblique and conical shock waves generated over wings and bodies.

Properties of strong planar shock waves as shown in Fig. 85, have been investigated extensively by Gross and his co-workers. It is of interest to note that this group has also obtained the fastest shock wave (collisional, 3600 kilometers per second or over 1% of the speed of light) in a shock tube, using deuterium at 1/20 millimeter of mercury or 7×10^{-6} atmoshpere, and 25°C. Here shock velocities are plotted for hydrogen, at an initial temperature of 273°K (note that at high temperatures the difference between the centigrade and Kelvin scales is not significant) and 0.1 mm mercury pressure (0.00013 atmosphere), from its sound speed value of 1284 meters per second (Mach number M of unity) right to the limit of the speed of light (300 million meters or one billion feet per second). By comparison, a shock from a TNT explosion starts in air with a maximum Mach number of M $\cong$ 20; re-entry from a satellite orbit in air occurs at M $\cong$ 25; from a nuclear explosive in air the shock starts at M $\cong$ 3000 (over 1 million meters per second). Other velocities (meteoroids, solar wind and solar-flare electrons and nucleons) are also noted for comparison.

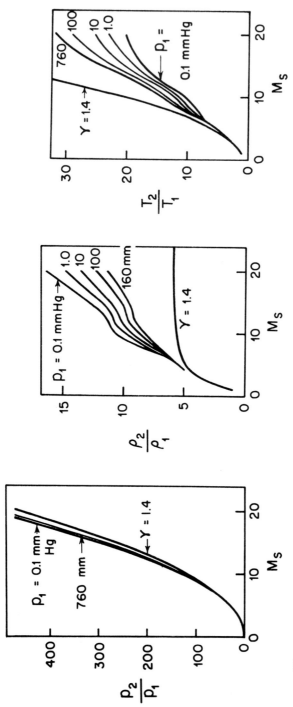

FIG. 84: CHARACTERISTICS OF PLANAR SHOCK WAVES

The variation of pressure (p_2/p_1), density (ρ_2/ρ_1), and temperature (T_2/T_1) ratios for planar shock-waves in air at an initial pressure p_1 are shown for an imperfect gas. The case for a perfect gas ($\gamma = 1.4$) with shock-wave Mach number M_s, is shown for comparison. This illustrates quantitatively what was shown qualitatively in Fig. 83. It is seen that the pressure ratio increases somewhat with decreasing initial p_1, from the perfect gas value, but significant increases in density and very large decreases in temperature occur. (Courtesy: UTIAS.)

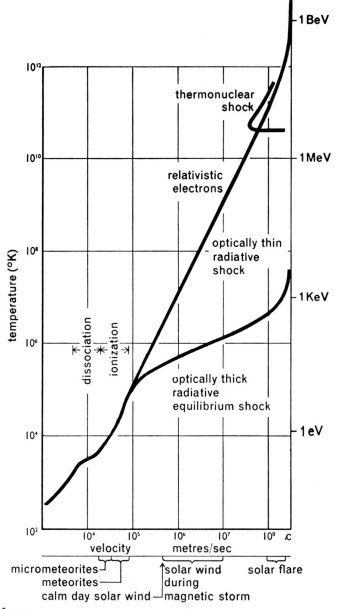

FIG. 85: SHOCK-WAVE TEMPERATURES FROM A FEW HUNDRED TO A FEW TRILLION DEGREES

The temperature behind normal shock waves in hydrogen at an initial pressure of 0.00013 atmospheres from the speed of sound to the speed of light. The temperatures are given in °K (C + 273°) and in electron volts (eV, prefix K = 10^3, M = 10^6, B = 10^9, leV ≡ 11,600°K). Some comparison velocities (solar flares, etc.) in meters per second are also shown. (Courtesy: R. A. Gross).

It is seen that hydrogen dissociation starts at a shock velocity of 5000 meters per second (M $\cong$ 4) and is complete at 20,000 meters per second (M $\cong$ 16). Full ionization is obtained at 80,000 meters per second (M $\cong$ 60) and the plasma temperature reaches nearly 100,000°K.

At 300,000 meters per second (M $\cong$ 230) the plasma has achieved a temperature of 1,000,000°K and the gas radiates strongly. At 10 million meters per second the temperature reaches 1 billion°K and the velocity of the relativistic electrons approaches 3 per cent of the speed of light. (The right scale gives the temperature in electron volts, 1 ev $\cong$ 11,600°K; 1 Kev $\cong$ 10⁷°K.) If the hydrogen contains its isotopes deuterium and tritium, then thermonuclear reactions will take place, depending on the product of the containment time, T, and the number density of the gas in particles per cubic centimeter, n. If the product n T is greater than 10^{15} and temperatures of 10^7°C (10 million°C) exist, then thermonuclear reactions are possible. Scientists and engineers have been working for nearly three decades on the problem of controlled thermonuclear fusion as a source of future power, but without success so far. When this problem is solved, the world's energy supply will be virtually unlimited. This matter becomes increasingly urgent as reservoirs of gas, coal and oil dwindle with the exponential rise in population and world industrialization.

Some Characteristics of TNT and Nuclear Explosions

Although the above stated conditions behind a plane or normal shock wave ideally remain unchanged, those for a spherical (or cylindrical) blast wave change drastically with time or distance from the explosion. Figure 86 shows what happens to the pressure profiles following a nuclear explosion. Initially, the peak pressure decays rapidly, inversely as the cube of the radius ($1/R^3$). As the shock approaches the strength of a sound pulse, the decay is more gentle, approximately inversely as the radius ($1/R$). The entire pressure profile changes behind the wave and even becomes negative (subatmospheric) as the shock wave engulfs an ever-increasing volume or mass of air as it races outward in time (1 to 4). At the shock front itself the planar values (Fig. 84) still apply. However, behind it, the quantities must be calculated with some difficulty, analytically or numerically.

The overpressure ($\Delta p = p_2/p_1 - 1$) in atmospheres across the shock front in an explosion, versus distance in feet is shown in Fig. 87 for a chemical explosive (one pound of TNT) and a nuclear explosive. (An equivalent point source of one pound of TNT was assumed. The term point source is used because virtually all of the energy is concentrated in a very small volume. See Chapter 5). A pound of TNT has a spherical radius of about 2 inches. At the point where the blast wave hits the air, the pressure behind it is about 440 atmospheres (although at the detonation-wave front, in the adjacent explosive gases it is about 160,000 atmospheres). For the TNT, the pressure initially decays inversely as the radius (1/R), followed by a more rapid decay, then finally at a rate approximately inversely as the radius once again, as the shock becomes a sound pulse. The point source, as noted above, initially decays as the inverse cube of the shock radius ($1/R^3$). The decay laws illustrate the difference between a very strong nuclear blast wave compared with that generated by a chemical explosive of finite size. The latter spreads the energy of the explosion over a greater initial driver-gas volume, thereby reducing the initial shock peak pressure and the decay rate.

At a shock radius of about 1 foot, the two curves cross and the point-source solution falls below the TNT solution and then runs almost parallel to it. That is, the history of the early part of the explosion has been "forgotten". The point source achieves very high initial temperatures (10^6°C). As a result, significant energy is bound up in the various excitation modes (ionization, dissociation, vibration) noted previously for an imperfect gas. This energy is not available to drive the shock wave. Therefore, in the weak shock wave region, at the same radius, the point source explosion has a lower overpressure, Δp, compared to TNT.

A similar graph for the detonation of a one-megaton nuclear bomb (surface burst) appears on Fig. 88. It shows the shock radius (R_s) and its arrival time t_s from ground zero as a function of overpressure Δp. For example, at 1000 t_s = 1.5 sec or t_s = 1.5 milliseconds, Δp = 10^5 psi (6800 atmospheres). Even a mile away ($R_s \cong 5300$ feet) from the point of blast the overpressure Δp, is greater than two atmospheres, a pressure that could produce devastations even beyond what is pictured in Figs. 31 and 32, since it would represent a fifty-fold increase over the TNT equivalent used on Hiroshima.

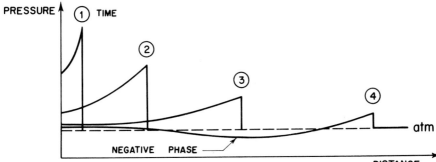

FIG. 86: PRESSURE PROFILES FROM A POINT-SOURCE EXPLOSION

Some typical pressure-distance curves for successive times after a point source explosion show the decay of pressure with distance and the development of the negative phase where the pressures are below atmosphere and the flow is in the negative direction (towards the origin of the explosion). (Courtesy: after G. F. Kinney).

Figures 87 and 88 are directly related by explosive "scaling laws". It can be shown that the characteristics of an explosion depend on the total energy, the density of the surroundings (medium) and the medium itself (which can be characterized by its specific heat ratio γ). For a given explosive the charge radius r_0, varies as the cube root of its mass. Therefore, 1000 pounds of TNT would develop tenfold the overpressure of a one-pound charge at the same radius. Alternatively, the same overpressure now will occur at tenfold the distance.

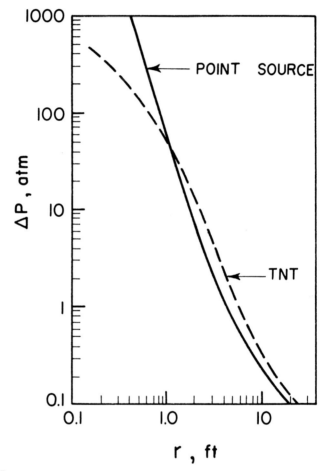

FIG. 87: COMPARISON OF TNT AND POINT SOURCE EXPLOSIONS

The peak overpressure ratio (ΔP) versus distance (r) are plotted for explosions from a spherical charge of TNT and nuclear source both having the same energy release of one pound of TNT. After a distance of 1 foot from the explosion, the decay rates are nearly the same. Near the explosion the point source decays rapidly, inversely as the cube of the radial distance, whereas the TNT charge decays more slowly, inversely with the radial distance. (Courtesy: after G.F. Kinney).

For example, from Fig. 87 read a 2.0 atmosphere overpressure at a 3.3 foot radius from a one-pound charge point source. Therefore, for a one-megaton surface explosion (a factor of 2 must be used to double the energy compared to a free-air burst) this overpressure should occur at $\sqrt[3]{2 \times 10^6 \times 2200} \times 3.3$ feet or 5400 feet. From Fig. 88, read about 5300 feet. This is close enough for graphical accuracy.

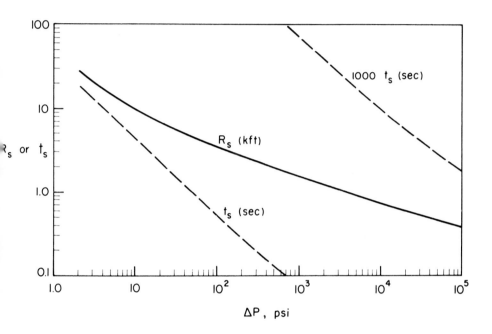

FIG. 88: SHOCK WAVES FROM A NUCLEAR BLAST

The shock radius (R_s) and shock arrival time (t_s) versus peak overpressure ΔP, for a 1-megaton surface burst. (Courtesy: after H. L. Brode).

7

OTHER TYPES OF SHOCK WAVES

In addition to the moving planar, cylindrical and spherical blast or shock waves, there are other types of shock waves that occur in supersonic (Mach number less than five, $M < 5$) or hypersonic flight (Mach number greater than five $M > 5$, where the shock wave system lies close to the body). Such wave systems can be put into a "steady frame" of reference by considering the aircraft at rest and the atmosphere rushing over it, just like a model in a wind tunnel. Such flow photographs are shown in Figs. 40 to 42, and are supplemented here with many other examples.

Oblique and Conical Shock Waves

Figure 89 shows two oblique shock waves produced by 14- and 21-degree wedges in a wind tunnel at $M = 2.5$. Knowing the flow Mach number and the wedge angles, the shock wave angles and all the other flow quantities can be found analytically and have been confirmed experimentally.

It can be seen that the oblique shock waves collide and give rise to a slipstream or contact surface C, which separates the gas compressed by the two right oblique shock waves (S_1, S_2) from the two on the left (S_3, S_4). It shows up as a discontinuous surface in the schlieren picture, since the density across it changes very rapidly (similar to a shock front). The pressure and flow direction are the same across it, but all other quantities are different. The third transition front that occurs in supersonic flow (in addition to the shock and contact surface) is the expansion wave (or its opposite type, the compression wave). It shows clearly as a triangular black patch with its apex at the corner of the lower wedge (white on the upper wedge). Such a system of wedges and waves might be used as inlet diffusers on future supersonic aircraft engines.

Figure 90 shows the oblique shock waves produced over a 10-degree wedge in a shock tube. The hot flow (2) processed by the shock wave is quite uniform, $M = 1.5$. The oblique shock waves strike the walls of the shock tube, interact with the boundary layers and reflect as expansion waves (dark wedge) embedded between two (light) compression-shock waves. This is an important type of interaction encountered in supersonic flight.

The cold flow (3) processed by the rarefaction wave through a torn diaphragm is eddying and rough. The shock waves and their interactions with the walls are therefore poorly defined. The flow resembles a turbulent flow in a pipe.

The infinite-fringe interferogram (a term used in interferometry when the entire test area before a run appears uniformly illuminated on the photographic plate; during the run the test area is covered by fringes, which in a two-dimensional flow are lines of constant density, as in this case) in Fig. 91 shows the flow F, over a 90-degree wedge-plate. The bow shock S, and the constant density contours are beautifully illustrated. The expansion fan radiates from the corner C. It reaccelerates the subsonic flow behind the bow shock S, to supersonic speeds over the flat plate at a Mach number slightly less than in

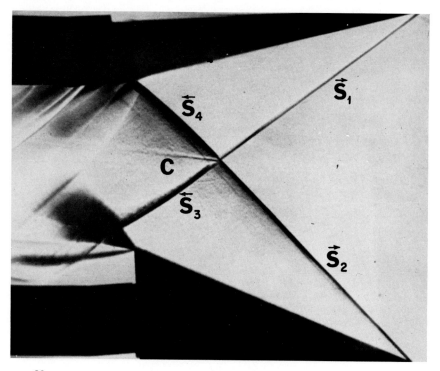

FIG. 89: INTERACTION OF OBLIQUE SHOCK WAVES

A schlieren photograph of the interaction of two incident oblique shock waves S_1 and S_2, in the UTIAS 16 × 16-inch supersonic wind tunnel, at a Mach number, M = 2.5, pressure, p = 0.003 atmospheres and a temperature T = −217 °F (−139°C). The upper wedge has an angle of 14 degrees and the lower 21 degrees. The slipstream (contact surface) C, after the interaction and the two deflected shock waves S_3 and S_4, are clearly seen (flow, right to left). (Courtesy: UTIAS).

the free-stream, F. Excellent quantitative density data can be obtained without disturbing the flow. For this reason interferometers are very valuable flow-diagnostic instruments.

The shadowgram of a blunted cone (Fig. 92) in flight at M = 3.4 in a

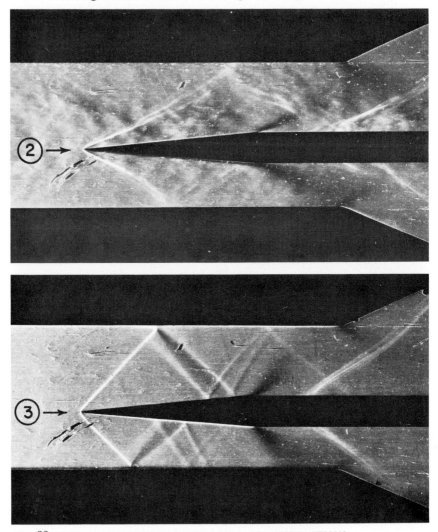

FIG. 90: OBLIQUE SHOCK-WAVE BOUNDARY-LAYER INTERACTIONS
Schlieren photographs show the flows and boundary layers in the uniform regions in a 2 × 7-inch shock tube. The (upper) flow in hot state (2) over a two-dimensional 10-degree wedge shows that it is quite uniform at a Mach number M = 1.50. The (lower) flow in cold state (3) at a Mach number M = 2.4, is rough and eddying. A clear but complex interaction of the oblique shock wave with the wall boundary layer appears in state (2) but not in state (3). The latter region is not suitable for testing. (Courtesy: UTIAS).

free-flight range (range tanks pressurized to 2 atmospheres to give high-contrast shadowgraphs) strikingly illustrates the bow shock, the expansion wave around the corner at the base of the projectile, the turbulent wake and the tail shock. Conical flows can be analysed numerically and solutions obtained for the flow properties. For sharp cones, the pressures are constant along conical lines, for example, the

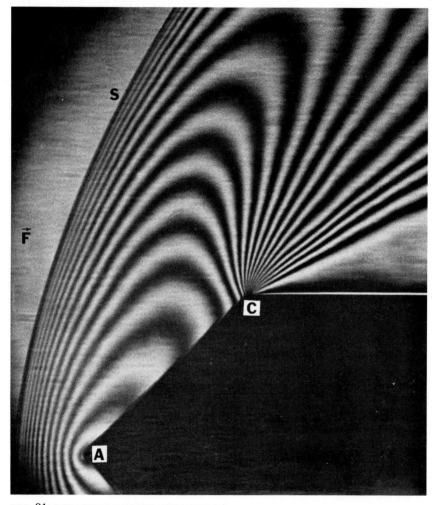

FIG. 91: DETACHED OBLIQUE SHOCK WAVE

The infinite-fringe interferogram shows an air flow F, at a Mach number M = 2.5 over a 90°-wedge plate in the UTIAS 16 × 16-inch Supersonic Wind Tunnel. The flow from the bow shock wave S, to the wedge apex A, and beyond to the corner of the flat plate is subsonic (M < 1). It expands around the corner C, to become supersonic again (M > 1) over the flat portion of the plate. (Courtesy: UTIAS).

cone surface itself. However, in all types of shock waves, their normal components are identical to those discussed earlier and their physical properties remain unchanged. It is the flow field *behind* the shock wave that changes with the different type of wave (planar, cylindrical, spherical, oblique, conical).

At very low speeds (subsonic) it is very difficult to significantly compress the air through which an aircraft is flying. As the speed increases from transonic to supersonic and finally to hypersonic Mach numbers the air becomes increasingly more compressible as more and more degrees of freedom become excited (Figs. 82 and 83). This fact is convincingly illustrated in Figs. 93 and 94. At a transonic

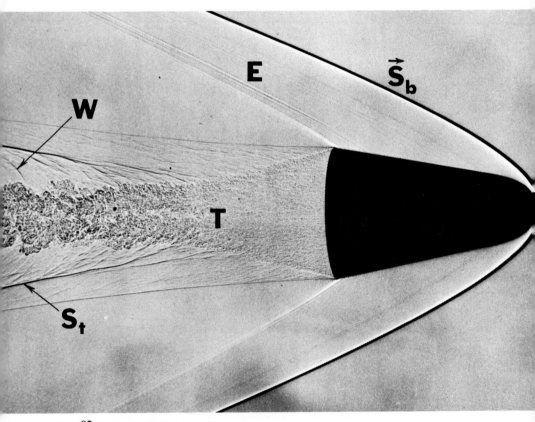

FIG. 92: CONICAL SHOCK WAVES

A shadowgram showing the flight of a 20-degree blunted steel cone at a Mach number M = 3.4, in air at 2 atmospheres in an aeroballistics range. The pressurized range provides excellent photographs of the bow shock wave S_b, the corner-expansion wave E, the tail shock wave S_t, the turbulent wake T, and the sound waves W, generated by the turbulence. (Courtesy: u.s. Naval Ordnance Laboratory).

Mach number M = 1.05 the bow shock S_b, is quite far from the sphere (1.11 body diameters), whereas at a hypersonic Mach number M = 17.5, it is very near the sphere (0.05 body diameters), or more than twentyfold closer.

Interaction of Shock Waves

Figures 95 and 96 have been included to further elaborate on the shock reflection and diffraction phenomena first noted in Fig. 80. In Fig. 95 a planar shock wave S_1, in a shock tube, moves past a half-plate, diffracts and also reflects as shock wave S_2.

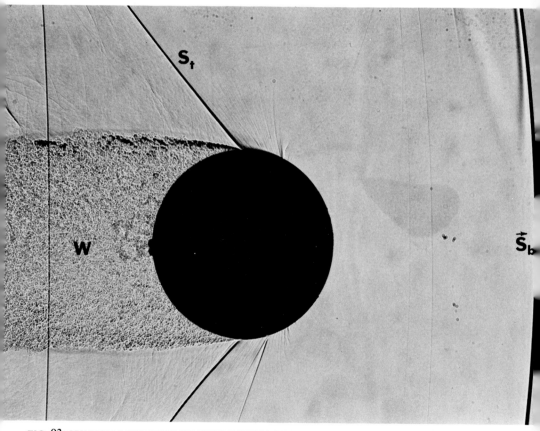

FIG. 93: SPHERICAL SHOCK WAVE AT TRANSONIC SPEEDS
Shadowgram of a 5.9-inch-diameter sphere in flight (to the right) through air in a ballistics range at a transonic Mach number M = 1.05. The boundary layer separates at the top of the sphere (90°) giving rise to tail shock waves S_t. The turbulent wake W, is very prominent. The bow wave S_b, is far (6.65 inches) from the sphere and is typical of transonic flow. (Courtesy: Naval Ordnance Laboratory).

This process gives rise to a contact surface C, which meets the original planar portion of S_1, and the curved diffracted part of S_1, as well as the curved reflected section of S_2, at the confluence of three shocks (a triple point). The subsonic flow accelerates over the top of the plate to become locally supersonic. In the same neighbourhood the shear flow over the plate generates a vortex V, which entrails the other end of the contact surface. It is the boundary line of the air

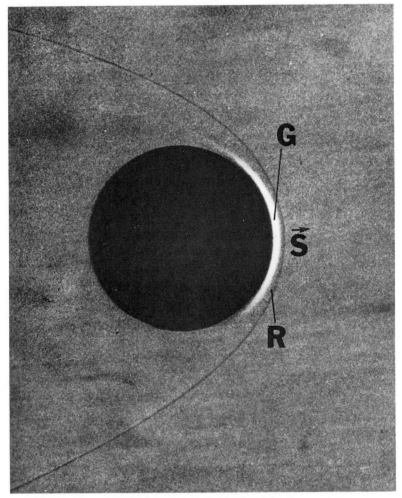

FIG. 94: SPHERICAL SHOCK WAVE AT HYPERSONIC SPEEDS

Schlieren photograph of a nylon sphere (3/16-inch diameter) flying (to the right) at a hypersonic Mach number $M = 17.5$, through xenon gas, in a ballistics range. Note how close the bow shock S_1 is to the sphere, and the relaxation distance R, between the shock and the radiating ionized gas, G. (Courtesy: Naval Ordnance Laboratory).

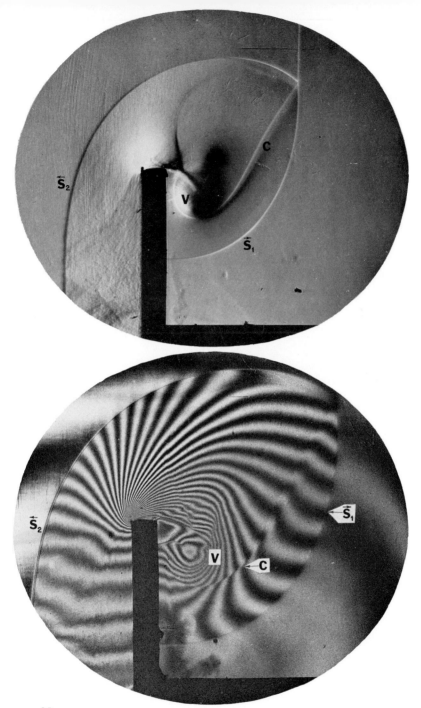

FIG. 95: SCHLIEREN PHOTOGRAPH AND INTERFEROGRAM OF SHOCK-WAVE REFLEC-
TION AND DIFFRACTION

The diffraction and reflection of a plane shock wave over a slender plate in a 2×7-inch shock tube shows the incident shock S_1, reflected shock S_2, vortex V, and contact surface (or slipstream) C. The flow behind the straight portion of S_1 is just supersonic ($M = 1.04$) and similarly behind S_2, it is brought to rest ($M = 0$). The air, initially at room temperature, had a pressure of 1/4 atmosphere. The schlieren photograph (upper) and interferograms (lower) show the same phenomena. From the latter it is possible to determine the density everywhere in the flow. (Courtesy: UTIAS).

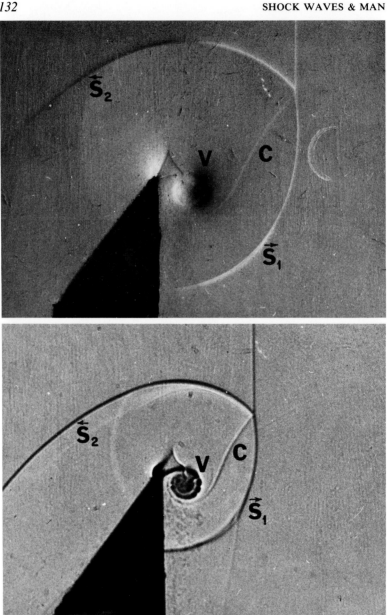

FIG. 96: SCHLIEREN PHOTOGRAPH AND SHADOWGRAM OF SHOCK WAVE REFLEC-
TION AND DIFFRACTION
The photographs, using a sharp-edged plate model, complement those shown in
Fig. 95. The roll-up of the spiral vortex is especially clear in the shadowgram and
illustrates how valuable optical methods are in the study of shock-wave
phenomena. (Courtesy: UTIAS).

heated and compressed by S_1 and S_2 that has come through from the other side of the plate. This process is illustrated by the very clear schlieren photograph (upper) and the fine infinite-fringe interferogram (lower). The fringes are contours of constant density. The sudden changes in density across S_1 and S_2, the mild density change across C, the expansion around the front corner of the plate (radial fringes) and the contours inside the vortex V, can be readily ascertained. Figure 96 is very similar to that of Fig. 95, with the exception

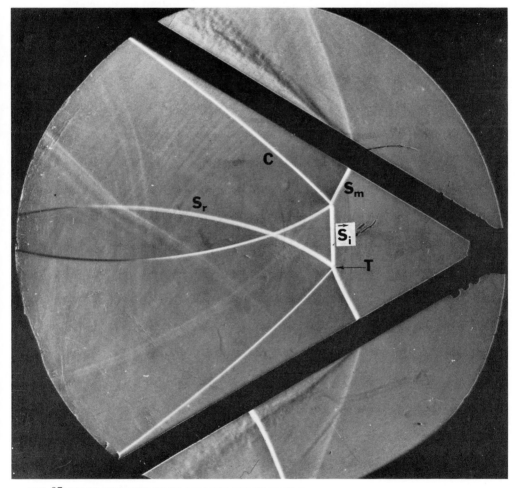

FIG. 97: MACH REFLECTION OF OBLIQUE SHOCK WAVES

A schlieren photograph of Mach reflection in a shock tube. The incident shock wave S_i, the reflected shock wave S_r, The Mach stem S_m, and the triple point T, where these three shock waves meet and the contact surface (slipstream) C starts, are clearly illustrated. This is a classic type of nonlinear reflection problem which is still not understood in all its details. (Courtesy: UTIAS).

that the top of the plate has a sharp wedge. The shadowgraph (lower) is particularly useful in showing the spiral structure of the vortex. The three optical methods illustrated in Figs. 95 and 96 show how important and complementary they are in examining compressible flows where changes in density and its rates of change can be detected and utilized for quantitative and qualitative studies.

Figure 97 is a schlieren photograph showing a planar shock wave moving into a converging duct undergoing Mach reflection at each wall surface. The wave elements (similar to Figs. 95 and 96) are symmetrical and very clearly defined. The triple point T, at the confluence of the three shock waves (S_i, S_r, S_m) and the contact surface C, is of particular theoretical interest in understanding the molecular collision processes that take place at this unique (or mathematically singular) point.

In order to illustrate a very practical application of shock-wave diffraction through a nozzle, Figs. 98 and 99 have been included. They show how the incident shock wave S_1, as it moves into a contoured supersonic nozzle, induces a uniform flow for aerodynamic-test purposes. To match the flow conditions in the nozzle requires the generation of a second shock wave S_2, and a contact surface C, (marking the boundary of the air which has come from the throat through the expansion waves E). Moments later S_1, S_2 and C are swept downstream (to the right) to give a uniform flow at the end of the nozzle. This process is illustrated by the schlieren photograph (upper) and interferogram (lower) of the flow. Figure 99 is an excellent complementary schlieren record of a flow in a different nozzle and with different starting conditions. The wave system is particularly clear. This type of starting process is fundamental to the so-called ''straight-through'' hypersonic shock tunnel, where very high Mach numbers with extreme stagnation temperatures can be achieved for short durations (fractions of a millisecond).

Shock Waves in Solids and Cratering Phenomena

It was shown that a rapid addition (deposition) of chemical or nuclear energy in a gas, liquid or solid causes a fireball of high-pressure, high-temperature gas which drives before it a blast wave or shock wave. Figures 14 and 15 illustrate the enormous craters that can be produced by an impacting meteorite, due to the gigantic quantity of mechanical (kinetic) energy that is released when a hypervelocity object is suddenly brought to rest. The shock formation and the cratering process in a solid can be illustrated clearly by firing a hypervelocity pellet (a 1/8-inch-diameter plastic sphere travelling at 21,800 feet per second) into a transparent plastic target, as shown in Fig. 100. In frame 1, the pellet P, is approaching the surface of the target T; in frame 2, the hemispherical shock wave S, generated on impact, is clearly evident and the vapourized plastic target and pellet are expanding backward as the cratering process C, proceeds; in

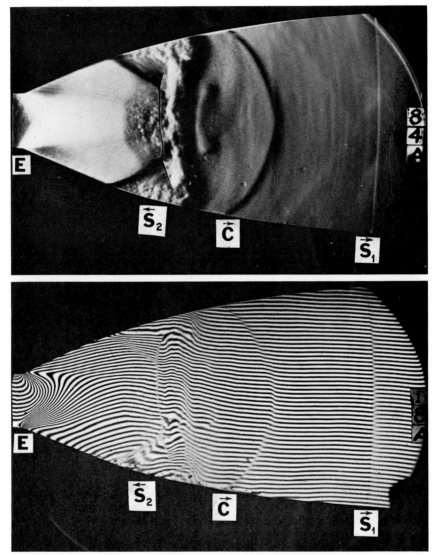

FIG. 98: DIFFRACTION OF A SHOCK WAVE THROUGH A SUPERSONIC NOZZLE

The interferogram (top) and schlieren photograph (below) show how a planar moving shock wave S_1, is used to induce a supersonic flow in a two-dimensional nozzle past the expansion waves E, at the throat, for short-duration (microseconds or milliseconds) aerodynamic tests. In the process of starting this flow, contact surface C, and second shock wave S_2, are also generated and swept away to give a uniform supersonic stream. The shock S_2, becomes bifurcated as it separates the nozzle and wall boundary layers. The schlieren photograph provides a good picture of the wave system, whereas the interferogram makes it possible to determine the density throught the nozzle flow. (Courtesy: UTIAS).

frame 3, the cratering process continues and the rear surface of the
target is seen to spall (fracture, F) as a result of the unloading tension
wave (rarefaction wave) that follows the reflection of the hemispherical shock wave from the rear surface.

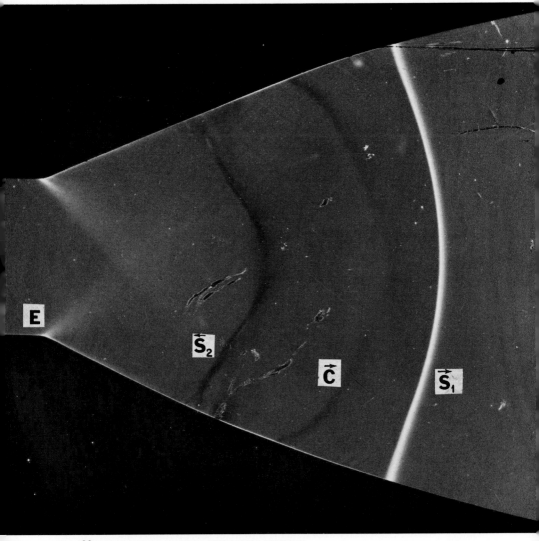

FIG. 99: SHOCK WAVE DIFFRACTION IN A NOZZLE

The schlieren photograph complements Fig. 98, for a different set of initial conditions.
The primary shock wave S_1 is very prominent. As one follows the flow from left to right,
the density decreases across the expansion waves E, increases across the secondary
shock wave S_2, and the contact surface C, and then decreases abruptly across the primary
shock wave S_1. (Courtesy: UTIAS).

Figure 101 shows similar craters produced in lead targets by different materials at different velocities. On the right is the crater produced by a 0.22-inch-diameter single-calibre (length equals diameter) plastic projectile P, at 4200 feet per second; in the centre, the velocity was raised to 14,300 feet per second; on the left, a high velocity copper fragment made the deepest crater. Relations exist that predict the crater-depth-to-diameter ratio as a function of hypervelocity and the material properties of the impact pellet and its target.

It has already been pointed out that micrometeoroid or meteoroid impact is a hazard to space travel on prolonged journeys. To cope with this problem, a meteoroid bumper has been suggested. Essentially it is a protective skin around the spacecraft. Its purpose is to vapourize the meteoroid into particles so small that when they strike the main skin of the craft little damage is done. Figure 102 shows the principle being tested in the laboratory. A 1/8-inch-diameter pyrex sphere S, is seen to approach a thin (1/16 inch) aluminum sheet (bumper, B) which is 1 foot from the main skin. When the pellet strikes, it vapourizes an area of the bumper and itself, and the gases expand backward and forward toward the main skin. The micrometeoroid droplets and solidified fragments strike the surface but are too small to do serious damage. It should be stressed that the

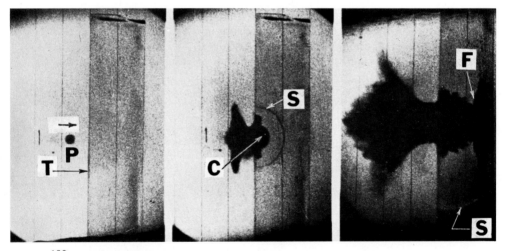

FIG. 100: SHOCK WAVES IN SOLIDS
A hypervelocity impact on a plastic target T, is shown by a 0.125-inch-diameter plastic sphere impacting at 21,800 feet per second. A framing rate of 1.4 million frames per second illustrates the event in three frames from left to right. Frame 1, before impact, shows the high-speed pellet approaching the front surface on the target. In frame 2, after impact, the crater C, and the shock wave S, in the solid plastic target are shown. The shock wave reflects as a rarefaction wave from the rear surface of the target. This causes a large tensile stress in the material and it spalls (fractures) F, as shown in frame 3. (Courtesy: USAF, AEDC).

molecular structure of shock waves in a gas are now well understood, whereas their structure in liquids and solids are still unknown.

It is also of interest to show that imprints similar to cratering can be caused by the very high gas pressures and temperatures of an implosion. Figure 103 shows, on the left, a copper disc with a small crater C_1, and one with a large crater C_2, and a separate side view V, where the copper has been pushed out. The small crater was caused by a hemispherical implosion driven by a detonable (explosive) mixture of 400 psi stoichiometric (two to one), hydrogen-oxygen whereas the large crater was produced when about 90 grams of PETN solid explosive was added as a hemispherical liner and exploded by the hydrogen-oxygen detonation wave (to be considered subsequently) to produce a powerful implosion. Under these conditions, the strength of the copper is negligible compared to the implosion pressures, and the metal flows like a liquid. Even if the implosion is permitted to go through a gun barrel it causes severe damage, as shown on the right of Fig. 103, at E and P.

Deflagration and Detonation Waves

The term "detonation wave" was just used (see also Fig. 79), and it is worth noting that in its simplest form it may be described as a planar shock wave (though in reality a complex cellular structure of Mach reflections) attached to a chemical reaction zone (where the heat of explosion is released and new gaseous products are formed), followed by an expansion wave. The structure is similar in explosive gases, liquids and solids. The shock wave is coupled to the chemical

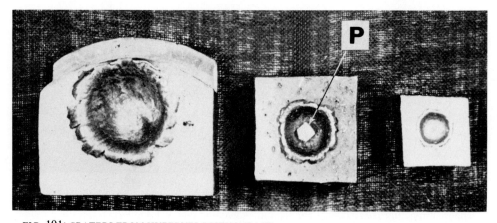

FIG. 101: CRATERS FROM HYPERVELOCITY IMPACT

A great deal has been learned about meteorite craters by using hypervelocity launchers. The three specimen lead targets show the progressively increasing crater size obtained from a 0.22-inch-diameter and 0.22-inch-long polyethylene plastic projectile P, impacting at 4,200 feet per second (right), 14,300 feet per second (centre) and from a copper fragment (left). (Courtesy: UTIAS).

process and its energy release. Consequently, for each set of material initial conditions (density and temperature) there is a unique detonation velocity. For example, for stoichiometric hydrogen-oxygen ($2H_2 + O_2$) the detonation velocities at 27 °C (300 °K) and 1, 10, and 100 atmospheres are 9310, 9730 and 10,100 feet per second, respectively; for TNT and PETN at a density of 1 gram per cubic centimeter, it is 16,450 feet per second and 18,200 feet per second, respectively. This remains unchanged whether it is a planar, cylindrical, or spherical detonation (explosion or implosion). In the latter case, very close to the implosion focus, the shock wave is decoupled from the chemical process and ideally races on to unlimited velocities as it approaches the point of collapse. In actuality, this process is limited by imperfect focussing, radiation and molecular collisions.

The formation of a detonation wave is illustrated in Fig. 104 in two sets of six photographs taken at intervals of a few microseconds. To the left, at 1.6 microseconds, a laser point-source ignites the detonable oxygen-acetylene mixture and, as time goes on, owing to the low energy of ignition, the shock wave separates from the flame front (reaction zone or deflagration wave) encompassing the burned gases. To the right, the energy of ignition is increased and the shock wave and the reaction zone stay together as a detonation front D. The

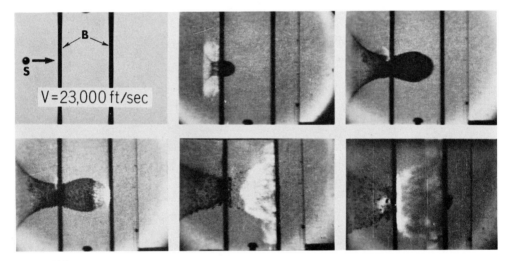

FIG. 102: METEOROID BUMPERS FOR SPACECRAFT
To guard spacecraft against possible meteoroid impact hazards, protective shields or bumpers have been devised. The series of photographs (top left to lower right), taken at 1.35 million frames per second, shows the approach of a 1/8-inch-diameter pyrex sphere S, at a hypervelocity of 23,000 feet per second, towards an aluminum bumper B, 1/16-inch thick, 1 foot away from a similar sheet assumed to be the main skin of the spacecraft. The explosive penetration and disintegration of the projectile and the preservation of the spacecraft are dramatically illustrated. (Courtesy: NASA).

mottled surface produced by spherical Mach wave reflections is clearly evident at 20.3 microseconds.

Figure 105 shows the structure of a detonation wave using interferometer, schlieren and soot techniques. The optical methods show up the complex topography of the Mach reflections, causing the detonation front to appear quite thick (by comparison, see Figs. 80 and 82 for schlieren and interferometer records of a normal shock front). The soot technique is used on the wall of the cylindrical detonation tube and when the triple points of the Mach reflections pass over the surface, the shearing action of a slipstream acts as a stylus and the criss-cross paths are traced. It is possible to modify and predict this pattern.

Shock Waves Produced by Focussed-Laser Radiation

Recent use of the laser for generating micro-explosions is of great importance in the field of the interaction of radiation with matter and its possible application to the production of thermonuclear fusion. The laser energies involved are presently about one thousand joules (1/4 gram of TNT) and one hundred thousand joules appear possible. Its deposition may occur ideally within the dimensions of an optical wavelength in extremely brief periods of 10^{-9} seconds (nanoseconds) or less, giving very high power outputs. Energy additions of 10^{17}

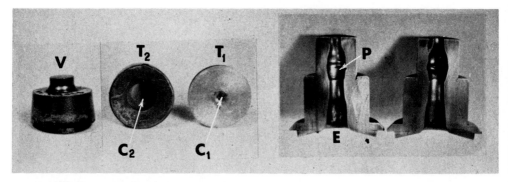

FIG. 103: CRATERING BY GASEOUS IMPLOSIONS

The craters left by gaseous implosions are very different from those produced by solid impact. In this case, the pressures and temperatures applied by the gas to a solid copper target are so great that its strength is negligible and the copper flows like a liquid. On the left are two targets (T_1, T_2) of 1.3-inch diameter. One has a small crater C_1, produced by an implosion originally driven by a two-to-one stoichiometric mixture of hydrogen-oxygen at 27 atmospheres. The larger crater C_2, was produced when 0.2 pounds of explosive PETN was added to the mixture. The extreme left view V, shows how the copper flowed into the gun barrel. On the right are two views of the severe erosion caused by an implosion moving through the entrance E, of a gun barrel (0.21-inch inside diameter $\times$ 1.25-inch outside diameter) and hitting a titanium projectile P, that was located 2 inches from the entrance (Courtesy: UTIAS).

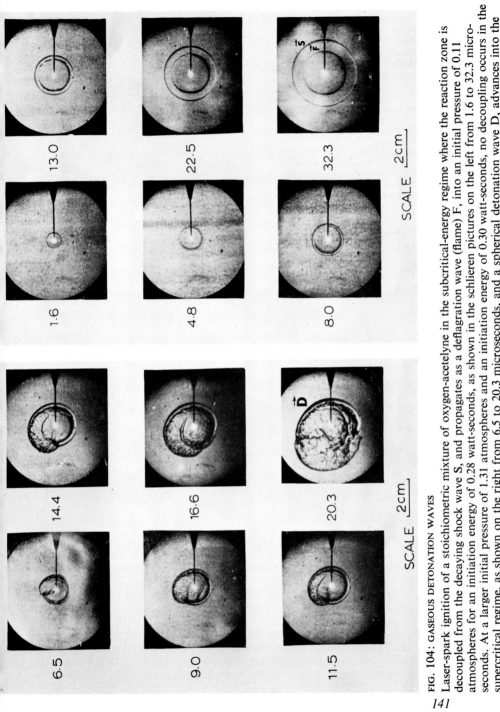

FIG. 104: GASEOUS DETONATION WAVES

Laser-spark ignition of a stoichiometric mixture of oxygen-acetylene in the subcritical-energy regime where the reaction zone is decoupled from the decaying shock wave S, and propagates as a deflagration wave (flame) F, into an initial pressure of 0.11 atmospheres for an initiation energy of 0.28 watt-seconds, as shown in the schlieren pictures on the left from 1.6 to 32.3 microseconds. At a larger initial pressure of 1.31 atmospheres and an initiation energy of 0.30 watt-seconds, no decoupling occurs in the supercritical regime, as shown on the right from 6.5 to 20.3 microseconds, and a spherical detonation wave D, advances into the mixture. (Courtesy: McGill University.)

141

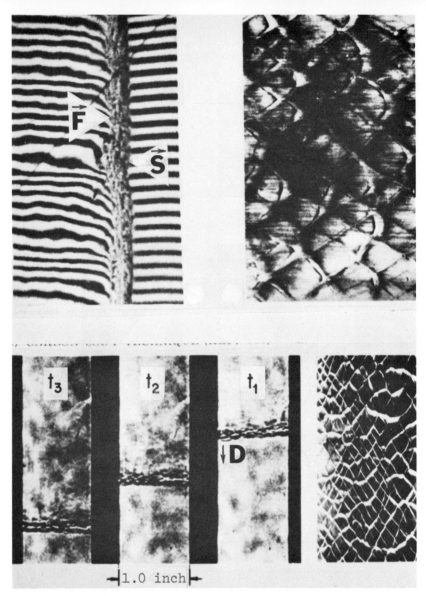

FIG. 105: OPTICAL AND SOOT TECHNIQUES FOR STUDYING DETONATION WAVES

An interferogram (upper left) shows clearly the structure of a planar detonation wave moving to the right in a two-to-one, hydrogen-oxygen, mixture. In front of the wave, the initial state is shown by the straight and parallel fringes. The shock wave S, followed by the flame front F, is thick and turbulent. The flow behind is reasonably uniform with a rising density. (Courtesy: General Electric). A soot (carbon black) pattern (upper right), left on the tube wall by a detonation, shows a diamond-like structure traced by the shearing action at the triple points of the Mach reflections of the detonation wave. (Courtesy: USSR.) Bottom row; three laser-schlieren photographs at 5-microsecond intervals, (t_1 to t_3) and a soot pattern (lower right) of a detonation D, in a four-to-one hydrogen-oxygen mixture. (Courtesy: University of California, Berkeley).

watts per square centimeter can be achieved. Great strides in the development and use of gasdynamic, chemical and other types of lasers will likely be sources of even greater energy deposition in the future.

A remarkable associated innovation is the employment of a second laser as a diagnostic tool to study the explosion produced by the main laser, having exposure times of 5×10^{-12} second intervals using schlieren optics, as shown in Fig. 106. Successive envelopes of the shock front are shown at exposure times of 5×10^{-12} seconds (5 picoseconds) at intervals of 5×10^{-9} seconds (5 nanoseconds). The shock waves take on a tear-drop shape, which is thought to be due to the fact that the explosion races towards the laser source, whose radiation couples with the ionized shock front to drive it faster, similar to a detonation wave which is coupled to its driving chemical energy. Since other theories can be utilized to interpret this process, a decisive explanation is not yet available. Nor is the pile-up of the plasma (white band) understood at present.

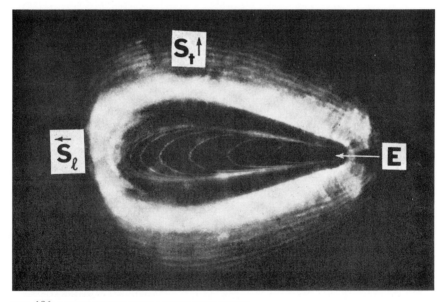

FIG. 106: LASER GENERATED MICRO-EXPLOSION
A schlieren record of an explosion E, in argon at 1 atmosphere produced by the discharge of a ruby laser. The explosion was photographed by using a second neodymium laser with a pulsed exposure time of 5 picoseconds (10^{-12} seconds) and a separation of 5 nanoseconds (10^{-9} seconds). The explosion energy was 1.6×10^{15} watts per square inch giving a longitudinal shock velocity S_l, of 6.6 million feet (1250 miles) per second at 5 nanoseconds, which then decayed to 0.3 million feet per second at 40 nanoseconds. The transverse velocity S_t decayed from 1.3 to 0.6 million feet per second in the same infinitesimal time interval. The major diameter is only 0.2 inches. (Courtesy: National Research Council of Canada).

8

SOME FACILITIES FOR STUDYING SHOCK-WAVE PHENOMENA IN THE LABORATORY

It is beyond the scope of this book to give more than a brief sampling of the facilities that are used today to investigate shock-wave phenomena (aside from nuclear and chemical explosives field tests). Five major pieces of equipment that have been developed over the past few dacades are outstanding. They are, the shock tube, the supersonic and hypersonic wind tunnels, the hypervelocity launcher and free-flight range, the low-density wind tunnel and the arc jet and its variant, the plasma tunnel.

Hypervelocity Shock Tube

The shock tube (see Chapter 6) is the "test tube" of high-temperature gasdynamics and physics. Numerous variations of this simple device exist in the form of combustion, electromagnetic, explosive, piston and simple high-pressure drivers that will burst a diaphragm to produce a planar shock wave in a channel which is followed by a region of high-velocity hot flow useful for aerodynamic and physical testing. This hot flow can be further expanded in hypersonic shock tunnel nozzles to produce high-temperature (enthalpy) and very high-Mach number (velocity) flows. Figure 107 shows a modern combustion-driven hypervelocity shock tube (with a 4×7-inch cross-section channel 48 feet long), at the University of Toronto, Institute for Aerospace Studies (UTIAS). It is capable of producing shock Mach numbers as high as 25 and temperatures of tens of thousand of degrees centigrade in a uniform test gas. Many interesting experiments have been performed in this facility on shock structure in argon, refractive indices of atomic oxygen and nitrogen and ionized argon, oblique shock-wave reflection, recombination rate coefficients of ionized argon, heat-transfer rates and ionized boundary layers in argon. These have provided new scientific data and an understanding of some of the physical processes encountered in flight at hyper-velocities.

FIG. 107: A MODERN SHOCK TUBE FACILITY

A view of the UTIAS Hypervelocity Shock Tube and ancillary equipment. The channel C, is 48 feet long, with a 4 × 7-inch cross-section. The combustion driver has a 6-inch internal diameter and a 14-inch external diameter. The channel, the dump tank D, a Mach-Zehnder interferometer I, a spectrograph S, and shock-wave pressure P, and velocity V, measuring equipment appear in the photograph. (Courtesy: UTIAS).

Shock Sphere

The shock sphere (the spherical counterpart of the shock tube) is a unique and simple facility for generating spherical and cylindrical shock waves in gases and liquids. It was conceived and developed at UTIAS. It is shown, together with its ancillary equipment, in Fig. 108. In this facility, glass spheres and cylinders were used as diaphragms (Fig. 67), and some photographs of explosions in air and water appear in Figs. 68, 71 to 73, and 76 to 78. These experiments provided new insight into low-energy explosion dynamics in gases and liquids that can have applications to nuclear-reactor safety, and the noise associated with blasting.

Explosively-Driven Implosion Shock Tube

Spherical and cylindrical implosion waves have been produced in the laboratory. Figure 109 shows a novel facility at UTIAS which utilizes explosively-driven implosion waves to drive intense planar shock waves (over 60,000 feet per second) in a 1.0-inch-diameter shock tube. Controlled, stable, focussed hemispherical implosions can be obtained on a routine basis, perhaps for the first time, in this unique apparatus. It has also been used for launching projectiles. Unfortunately, the pressures (millions of atmospheres or megabars), densitites (grams per cubic centimeter) and temperatures (tens of thousands of degrees centigrade) produced by the implosion and its reflection are so great that the projectiles disintegrate beyond velocities of about 15,000 feet per second. It is an excellent and safe facility for producing uniform flows of a few microseconds in duration. Significant results were obtained recently on radiative relaxation behind strong shock waves in air.

Supersonic Wind Tunnel

The postwar years brought with them very significant increases in size and speed of transport aircraft. To accomplish these impressive developments, advances in many disciplines were required including a great deal of testing. The NAE (National Aeronautical Establishment, Ottawa, Canada) large, modern, trisonic (subsonic, transonic, supersonic) wind tunnel, shown in Fig. 110, is representative of the type of facility that is used today for high-speed-flow tests. It is a blowdown (short-duration runs) type of tunnel with running times from 10 to 30 seconds with a low stagnation temperature of 300°K, representative of low supersonic Mach numbers. The tunnel Mach number range from 0.15 to 4.25 spans the trisonic regions. The maximum Reynolds number (a nondimensional parameter relating ineria to viscous forces in a flow, which is of importance in scaling model tests to full size) is about 50 million. It is usually achieved in testing two-dimensional high-speed airfoils (wings). The tunnel has now been in operation for a decade, serving the aerospace industry in Canada and elsewhere.

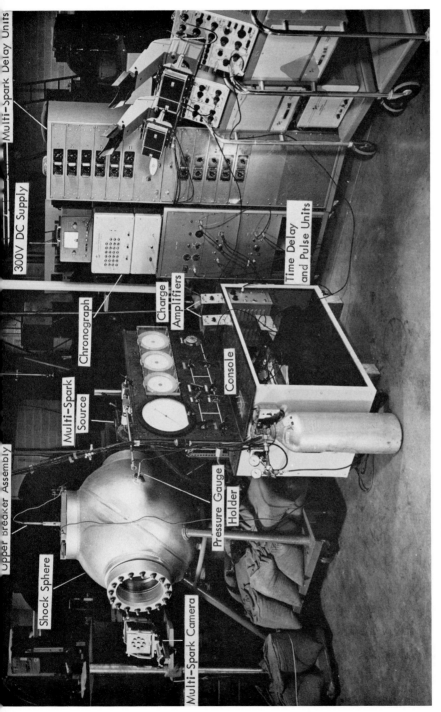

FIG. 108: FACILITY FOR STUDYING SPHERICAL AND CYLINDRICAL EXPLOSIONS

The Shock-Sphere facility and the ancillary apparatus used for generating cylindrical and spherical explosions in air and water. (Courtesy: UTIAS.)

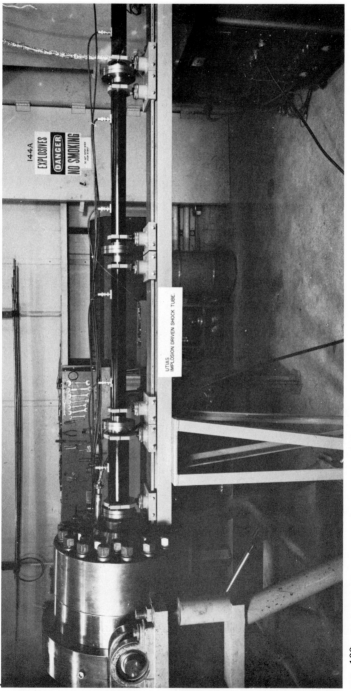

FIG. 109: AN IMPLOSION DRIVER FOR A LAUNCHER AND A SHOCK TUBE

The UTIAS Implosion-Driven Shock Tube (right) showing the explosive-driven implosion chamber (left) and some ancillary equipment. (Courtesy: UTIAS.)

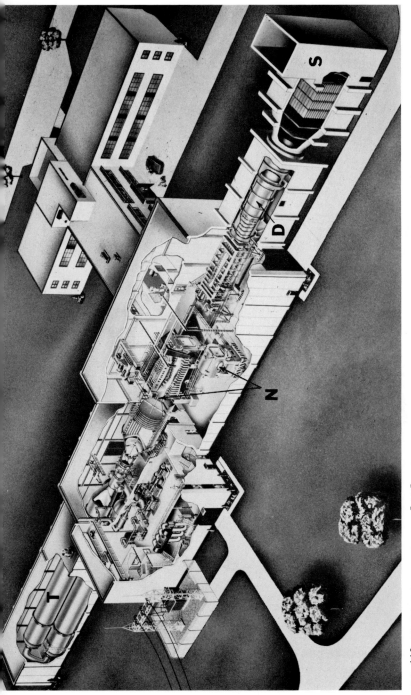

FIG. 110. SCHEMATIC VIEW OF NAE 5 × 5-FOOT TRISONIC WIND TUNNEL

This unique facility has been in operation for a decade assisting industry in the development of high-speed aircraft. Air from the high-pressure storage tanks T, blows through the nozzle-test-section N, to the diffusers D, and into the noise silencers S. Runs of 10- to 20-second duration can be obtained. (Courtesy: NAE and Dilworth, Secord, Meagher and Associates Ltd.)

149

Hypersonic Shock Tunnel

Figure 111 shows an early hypersonic shock tunnel (Mark I), erected at UTIAS using a test section obtained from the Cornell Aeronautical Laboratory, Buffalo. The driving shock tube (2×2-inch) is coupled to a nozzle section, test section and dump tanks. The reflected shock wave generates a pocket of high-temperature high-pressure gas which is expanded in the nozzle to high Mach numbers ($M \cong 15$). This facility has since been replaced by one of more modern design, where the beautiful laser-schlieren photographs shown in Figs. 42 and 46 were obtained.

FIG. 111 : HYPERSONIC SHOCK TUNNEL FACILITY

A view of the UTIAS Hypersonic Shock Tunnel, Mark I, and auxiliary equipment. (Courtesy: UTIAS).

Low-Density Tunnel

Low-density wind tunnels operate in the high-altitude pressure range equivalent to about a millionth (10^{-6}) part of an atmosphere which exists at about 300,000 feet, where the mean-free path (average distance between molecular collisions) is about one inch. Consequently, it is possible to study the aerodynamics of shapes which are now smaller (whereas normally the body is many million times larger) than the mean free path. That is, the ratio of this path to body diameter, known as the Knudsen number, is greater than unity. Under such circumstances the model drag and heat transfer properties depend on the individual collisions with the body (free-molecule flow) and the type of reflection that the molecules have to undergo (specular with a preferred direction; diffuse, with none) and their accommodation to the surface temperatures. In addition, since the operation is at very low pressure (density) a shock wave can be magnified a millionfold (say by reducing the pressure to a millionth of an atmosphere) and its thickness increases from a microinch to an inch. This makes it possible to study shock-wave structure in some detail with finite size (say, 1/16-inch diameter) pressure or temperature probes. A great deal has been learned from just such tests at UTIAS and other laboratories.

A schematic view of the UTIAS low-density wind tunnel appears in Fig. 112. The mechanical and oil diffusion vacuum pumps remove the gas from the surge chamber and test section leaving them at very low pressures. The flow meter admits a predetermined amount of gas into the stagnation chamber. The gas is then expanded in a free jet in the test section, where models are inserted and flow measurements are made.

Plasma Tunnel

The high-powered electric-arc jet or plasma jet has played an important role in the study of ablating materials for re-entry capsule heat-shield designs. It is equally useful for studying the aerodynamics and heat-transfer of planetary-entry models at high Mach numbers and large stagnation temperatures. A modern, unique facility of this type is shown in Fig. 113. It has a static pump-down pressure of about 10^{-8} atmospheres (3×10^{-5} millimeters of mercury). Such a high vacuum is obtained by literally freezing the gas molecules on the liquid-neon and liquid-nitrogen cooled freeze-out surfaces. Back-up mechanical and diffusion pumps are also used.

Helium, argon, krypton, nitrogen or other gases enter the source chamber from a plasma jet (high-power arc with up to 10% ionization), a radio frequency source (up to 1% ionization) or an ion beam which is neutralized (up to 95% ionization). From there the gas is expanded in a free jet to flow over the model under test. An electron gun or other instruments are then used to probe the flow, as illustrated in Fig. 113. It is possible to operate the tunnel in the Mach number

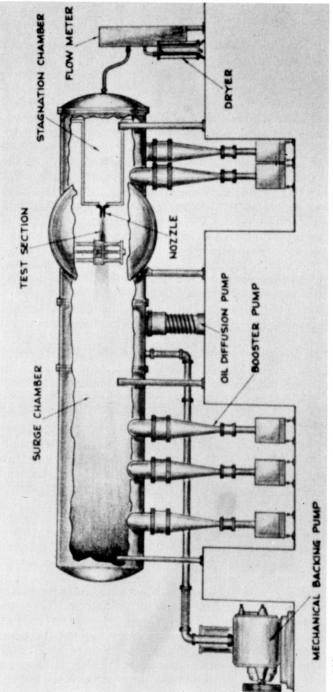

FIG. 112: UTIAS LOW DENSITY WIND TUNNEL

The schematic diagram illustrates the essential components of a modern low density wind tunnel. It can be pumped down to a pressure of 1×10^{-8} atmospheres (10^{-5} millimeters of mercury), which is equivalent to an altitude of 400,000 feet). The range of test Mach numbers is from 1 to 15, with a range of molecular mean free paths from millimeters to centimeters under actual experimental conditions. (Courtesy: UTIAS.)

range, $1 < M < 40$; velocity range, 300 m/sec $< V < 150$ Km/sec; total temperature range $300°K < T_0 < 8500°K$ and a mass flow range of 0.1 mg/sec $< m < 1$ g/sec (1/10 milligram to 1 gram per second).

Hypervelocity Free-Flight Range

Free-flight (open ballistic) ranges have been used for many years for testing armaments travelling at supersonic speeds. However, with the advent of ballistic missiles, supersonic fighters, bombers, and transports, the meteoroid hazard during space travel, and atmospheric re-entry, the closed hypervelocity free-flight range became a major test facility. Unlike the open air range, the hyperballistic range can be evacuated or pressurized with any desirable test gas. Since

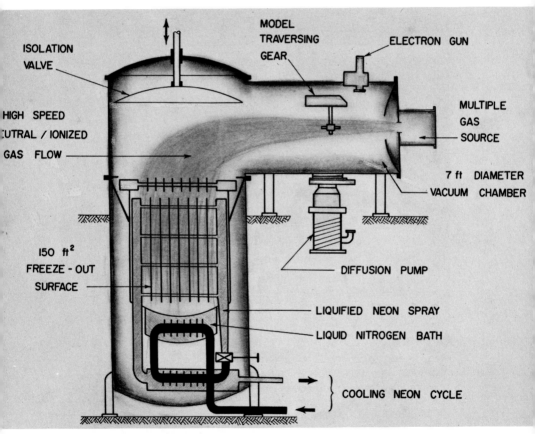

FIG. 113: UTIAS LOW-DENSITY PLASMA TUNNEL

This versatile piece of equipment is capable of flow velocities up to 150 kilometers per second, total temperatures up to 8500°K, and mean free paths larger than the dimensions of the facility. This makes it possible to study a wide variety of flows over large ranges of Mach number and Knudsen number. (Courtesy: UTIAS).

FIG. 114: HYPERVELOCITY FREE-FLIGHT RANGE

The upper photograph shows a 50-foot launcher (gun) consisting of a propellant powder chamber and piston compression tube followed by a disposable-liner launch tube using a light-gas (hydrogen) driver. Estimated maximum operating pressures of 350,000 psi (over 20,000 atmospheres) made it possible to launch models into the evacuated firing range (shown below) at 30,000 feet per second (greater than an Earth's satellite circular orbit speed of 26,000 feet per second). The left portion of the range is shown instrumented for studying the flash of light on impact of a projectile with a target (meteoroid-impact flash). (Courtesy: Computing Devices of Canada).

conventional guns are limited by the heavy explosive driver gases to three or four thousand feet per second, several sophisticated types of light-gas guns, driven by high-temperature high-pressure hydrogen or helium, were developed over the last two decades that could fire 0.22-inch-diameter plastic projectiles weighing a fraction of a gram to hypervelocities of 30,000 feet per second. Specialized explosive-driven launchers that were destroyed during each run accelerated a 5/8-inch-diameter half-calibre projectile of 0.06 ounces (2 grams) to 40,000 feet per second (by comparison the thin-rod sound speed in aluminum is 17,000 feet per second). This is still far short of the assumed maximum expected meteorid impact velocity of 240,000 feet per second (see Chapter 2).

Figure 114 illustrates the gun and range used by Computing Devices of Canada to produce projectile velocities of 30,000 feet per second. With the cutbacks in the space program this facility has unfortunately been closed. However, a number of open and closed free-flight ranges are still in operation at the Defence Research Establishment, Valcartier, Quebec, Canada, where a great deal of research and development work was conducted over the past two decades in the areas of ballistics and hypervelocity.

Sonic-Boom Simulator

A very recent application of the shock tube has been in the production of weak rather than strong shock waves in a facility known as a travelling-wave sonic-boom simulator. The era of the supersonic fighter and bomber is already here, and that of the supersonic transport is under way. Consequently, the sonic booms (see Chapter 3) illustrated in Fig 38 will be felt by millions of people throughout the world.

To understand the human, animal and structural response to sonic boom, a shock-tube driven and mass-flow-valve-driven travelling-wave facility was built at UTIAS. A major component is a concrete horizontal pyramid 80 feet long with a 10 by 10-foot base closed by a wave-reflection-absorbing, recoiling piston. The concrete horn, enclosed at both ends by two laboratories containing ancillary equipment, is shown in Fig. 115.

The N-wave from the tapered shock tube (Fig. 116 c) was generated by a diaphragm pressure ratio of 1.7. It has a very sharp rise time (50 microseconds) and a short duration (nearly 2 milliseconds). The N-wave produced by the mass-flow valve (d) originated from a reservoir tank pressure ratio of 1.7. It has a longer rise time (400 microseconds) and a much longer duration (over 80 milliseconds). The boom with the short rise time is very useful for studying startle effects on humans and animals arising from the sharp crack produced by such a boom, whereas the longer-duration boom with a greater impulse is more appropriate for investigating structural response.

FIG. 115: A SONIC BOOM SIMULATOR

A view of the UTIAS Travelling-Wave, Sonic-Boom simulator (upper view) showing the pyramidal, concrete horn, 80 feet long with an open end 10 × 10 feet (housed on the left, lower view). The sonic booms are generated by a shock-tube or a mass-flow-valve driver. A loud-speaker-driven booth for generating booms and other sounds for psychoacoustic studies is housed in the covered structure on the right, which also contains the shock tube, the mass-flow valve and ancillary equipment. The housing on the left contains the boom reflection-absorbing piston and equipment for studying human, animal and structural response to sonic booms. (Courtesy: UTIAS).

Recent studies at UTIAS have shown that the type of sonic boom that can be expected from overflights have little effect on increased heart rate and its subsidence. No more, in fact than being asked a simple question such as : "what is 7×7?" It has also been found that the effect on hearing over a 24-hour period is insignificant. Qualitative effects have been noted on tasks similar to driving an automobile, but this is now being researched to determine quantitatively if the booms might have any bearing on accident proneness. Work is also underway on quantifying the effects of sonic booms on plaster walls to verify the truth of crack-damage claims. Experiments are being conducted with small shock tubes in the field to determine animal and bird response to sonic boom. This is of particular importance for scheduling of supersonic flights over northern Canada to ensure that wild animal herds will not stampede as a result of the sudden startle from a sonic boom or that eggs will not be smashed by the reckless flight of startled nesting birds. Many additional experimental and analytical problems are planned for study in the near future. It is hoped that the results will make it possible for Parliament to pass reasonable flight regulations based on an understanding of the effects of sonic boom on humans, animals and structures.

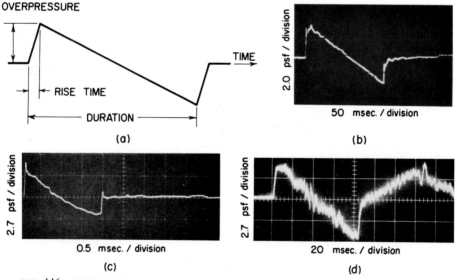

FIG. 116: ACTUAL AND SIMULATED SONIC BOOMS

The sketch (a) defines an ideal N-wave. The record in (b) shows an actual sonic boom produced by the Concorde SST, with a total duration of 260 milliseconds and maximum overpressure of 4 pounds per square foot (see horizontal and vertical scales). The records (c) and (d) depict sonic booms produced in the UTIAS Travelling-Wave Simulator driven by a shock tube and a mass-flow valve, respectively. (Courtesy: UTIAS).

9

SHOCK WAVES AND THE HUMAN CONDITION

The scientific and technological study of cosmic and terrestrial shock waves can be fascinating, creative and productive. Unfortunately, it is also bound up with a threat to human life from firearms, chemical explosives, and nuclear weapons on a scale that extends from individual tragedy to a possible world catastrophe involving millions of people. Hopefully, this realization will ultimately abolish aggression and war and lead to the evolution of a constructive and peaceful community of Man.

Shock Waves and Survial

The foregoing overview of shock-wave phenomena in Space and on Earth has illustrated their universal importance. We are just now beginning to understand their cosmic roles. It is doubtful that life on Earth would be possible if shock waves generated by nuclear reactions did not help to keep the solar furnace in operation. We do not yet really understand the full meaning of exploding stars and galaxies in the scheme of the cosmos and their filling of the vast spaces with rarefied plasmas and electromagnetic fields. Even our own solar system may owe its existence to some such cosmic explosion whose shock wave scattered the elements from which our Earth and, eventually, its innumerable life forms, evolved.

On Earth, man is at the mercy of shock waves and associated potentially dangerous phenomena generated by thunderstorms, earthquakes, volcanic eruptions, meteoroid impacts, and explosions in mines and elsewhere. These have taken countless lives and will doubtless continue to exact their toll in the future. The losses will be reduced when we have learned to recognize, monitor and measure, in advance, possible warning signals (pressure waves, temperature distributions, gas concentration, Earth strains, etc.) of impending disasters. The necessary steps can then be taken to minimize the loss of life and property. In Japan, for example, improved building design to cope with earthquakes, and the rapid marshalling of civil and military safety squads, are part of an effective initial effort to reduce casualties. Satellite photographs of impending hurricanes have already helped to save thousands of lives in Florida by providing sufficient advance warning to allow for the timely evacuation of people. In the future, methods may be developed to provide the same advance notice of volcanic eruptions and earthquakes. Fortunately, meteoroid impact is not common enought to be considered a great hazard, but there is always the possibility that a great disaster may occur. (Imagine the consequences had the Tunguska meteor of 1908 exploded over a populated region). Explosions in mines and elsewhere could be minimized by the enforcement of more stringent safety precautions.

Yet the loss of life from these natural causes is dwarfed when compared to the mindless slaughter of combatants and innocents alike that has taken place periodically since the dawn of our existence and especially in the wars and concentration camps of recent times. Admittedly, the invention of chemical explosives followed by nuclear weapons has increased the efficiency of killing. However, this is only one side of the coin, even if we discount the enormous possible uses of chemical and nuclear explosives for peaceful applications. It is a fact that the invention of nuclear (as well as biological and chemical) weapons, by making possible the annihilation of the species, has for the time being actually deterred the superpowers from going to war. They know full well that their use would mean mutual suicide and perhaps an end to most higher life forms on Earth. Even so, we cannot live indefinitely under this constant threat of annihilation and ultimately, all atomic, biological and chemical (ABC) weapons must be dismantled and their deadly charges immobilized or used for peaceful purposes, and war itself must somehow be abolished, if we are to survive on this planet. It is well to remember that at the time when these devices of destruction were being developed and were proliferating, some of the most heinous and brutal mass murders of all time were committed in extermination camps operated by the most developed of nations, without the use of any explosives at all. Even in the so-called underdeveloped nations, that primitive tool of civilization, the knife, has taken a staggering toll in the last few decades.

Such consuming brutalities, fanned by wild ideas and flaming words, dehumanize and precipitate acts of frightful destruction. The abolition of causeless hatred through education and law is as vital as the renunciation of violence and war. The humanities and the sciences, therefore, have a joint and inseparable responsibility for the maintenance of the unique and precious life on spaceship Earth. For it is doubtful if among the myriad of planets in the cosmos we will ever find a duplicate of our world.

Shock Waves and Man

The study of shock waves was started only recently by a group of applied mathematicians and experimenters who worked about the middle of the nineteenth century (Stokes, Earnshaw, Riemann, Rankine, Hugoniot, Vieille, Chapman, Jouguet and others). During World War II, and up to the present time, new and precise data have been required on blast-wave effects from chemical and nuclear explosives and shock-wave effects generated by re-entering missiles and supersonic aircraft. This need has provided a renewed impetus to many studies by some outstanding scientists (Taylor, von Neumann, Sedov, Bethe, Teller, Bleakney, Laporte, Kantrowitz, Patterson and others). A vast literature has been published which contains the considerable knowledge accumulated in this field during the past few decades.

Let us, finally, reflect for a moment. Can we say that we have learned as much about man himself after millennia of observation and countless commentaries? Ultimately it is the individual who has the power to decide how the knife shall be used: to sustain life, or to take life. Many lofty and inspiring religious, philosphical, economic and political systems, utopias and civilizations have been devised, or have evolved, to help man improve his lot spiritually and materially. Undoubtedly, they have had a civilizing influence on him which has been oscillatingly progressive and regressive. At best, however, their success has been only limited if we are to compare his bloody past history with his present agonizingly precarious condition and the questionable quality of his life. Can it be that our limited success in uplifting and civilizing our species stems from the fact that we have been describing (and prescribing for) man as we may perhaps *wish* him to be, rather than as he *really* is? (Or as Dennis Gabor has recently expressed it: "I *believe* in the perfectibility of man, because this is the only working hypothesis for any decent and responsible man. But I *know* of the almost infinite corruptibility of man"). What then is he?

This ageless question was asked very poignantly by the Psalmist some millennia ago:

> *"When I behold Thy heavens, the work of Thy fingers,*
> *The moon and the stars, which Thou hast established;*
> *What is man, that Thou art mindful of him?*
> *And the son of man, that Thou thinkest of him?*
> *Yet Thou hast made him but little less than God,*
> *And hast crowned him with glory and honour."*

He asked this question and paid this magnificent tribute to man the creator despite the Psalmist's knowledge of the earlier dialogue between God and Cain, which sheds an incisive light on man the destroyer:

> *"And Cain spoke unto Abel his brother,*
> *And it came to pass, when they were in the field,*
> *That Cain rose up against Abel his brother*
> *and slew him. And the Lord said unto Cain:*
> *'Where is Abel thy brother?' And he said:*
> *'I know not; am I my brother's keeper?' "*.

Each of us; our senate and legislative committees, our judicial bodies and inquiries and our courts, have wrestled with these problems and will undoubtedly continue to do so in the future. Only when we have some convincing answers and are able to accept ourselves as we are, can we hope to channel our destructive drives for power and pleasure into the constructive and civilizing pursuit of being our "brother's keeper". Then shock waves created by man, like his primitive knife, will ultimately be used only to sustain life and elevate man's material and spiritual lot to new, undreamed of heights.

CHAPTER REFERENCES

Chapter 2

Thunder

HAMMOND, A. L. Aug., 1973. Hurricane Prediction and Control. Impact of Large Computers. *Science*, pp. 643–45.

HILL, R. D. 1972. Thunderbolts. *Endeavour*. Vol. 31, No. 112, pp. 3–9.

MALAN, D. J. 1963. *Physics of Lightning*. London: English Universities Press.

PENNER, S. S. Oct., 1972. Elementary Considerations of the Fluid Mechanics of Tornadoes and Hurricanes. *Astronautica Acta*, pp. 351–61.

PLOOSTER, M. N. 1971. Numerical Model of the Return Stroke of the Lightning Discharge. *Physics of Fluids*, Vol. 14, No. 10, pp. 2124–33.

REMILLARD, W. J. 1960. *The Acoustics of Thunder*. Harvard University, Technical Memorandum No. 44.

RIBNER, H. S. and F. LAM. 1972. Acoustics of Thunder. Presented at 83rd Meeting, *Acoust. Soc. Amer.*, Buffalo, 18–21 April; July, 1972. Abstract in *J. Acoust. Soc. Amer.*, *52*, no. 1 (Part 1), p. 115. (Further developments by Leung, K. A. and D. Kurtz, unpublished.)

SIMPSON, R. H. Sept., 1973. Hurricane Prediction: Progress and Problem Areas. *Science*, pp. 899–907.

Earthquakes

ABELSON, P. H. May, 1973. Observing and Predicting Earthquakes. *Science*, p. 819.

ALFORS, J. T. et al. 1973. *Urban Geology–Master Plan for California.* California Division of Mines and Geology, Bulletin 198, Sacramento, California.

ALPAN, I. Dec., 1969. Earthquakes and the Engineer. *Technion Magazine, Israel Institute of Technology, Haifa*.

BATH, M. Nov.-Dec., 1968. From Lisbon to the Moon, *Scientia*, Vol. C111, No. DCLXXIX-DCLXXX, pp. 1–7.

EPSTEIN, H. M. and R. W. KLINGENSMITH, 1964. Detection and Identification of Nuclear Bursts. *Battelle Technical Journal*, Vol. 13, No. 8, pp. 3–9.

EVERNDEN, J. F. 1971. Discrimination Between Small-Magnitude Earthquakes and Explosions. *Journal of Geophysical Research*, Vol. 76, No. 32, pp. 8042–55.

HAMMOND, A. L. May, 1973. Earthquake Predictions: Breakthrough in Theoretical Insight? *Science*, pp. 851–53.

—— June, 1973. Earthquake Predictions (II): Prototype Instrumental Networks. *Science,* pp. 940–41.

HODGSON, J. H. 1964. *Earthquakes and Earth Structure*. Toronto: Prentice-Hall. See also 1966. Seismology and Earthquake Engineering. *Impact*. Vol. XVI, No. 4, pp. 297–320.

LEET, L. D. 1970. Earthquakes. *Encyclopedia Americana*, New York: Americana Corporation. Vol. 9, pp. 544–51.

MYERS, H. R. Jan. 11, 1972. Extending the Nuclear Test Ban. *Scientific American*, pp. 13–23. Vol. 226, No. 1.

SARION, S. et al. 1971. Excitation of Seismic Surface Waves with Periods of 15 and 70 seconds for Earthquake and Underground Explosions. *Journal of Geophysical Research*, Vol. 76, No. 32, pp. 8003–20.

SCHOLZ, C. H., L. R. SYKES, and Y. P. AGGARWAL, Aug. 1973. Earthquake Prediction: A Physical Basis. *Science,* pp. 803–10.

TOLSTOY, J. *The Pulse of a Planet*. 1971. New American Library. New York.

WHITCOMB, J. H., J. D. GARMANY and D. L. ANDERSON. May 1973. Earthquake Prediction: Variation of Seismic Velocities before the San Francisco Earthquake. *Science*, pp. 632–35.

WICK, G. Mar., 1972. Nuclear Explosion Seismology: Improvement in Detection. *Science*, pp. 1095–97.

Volcanic Eruptions

GROVE, N. July, 1973. A Village Fights for Its Life, *National Geographic Magazine*. pp. 40–67. Vol. 144, No. 1.

SYMONS, G. J. ed. 1888. The Eruption of Krakatoa and Subsequent Phenomena. *Report of the Krakatoa Committee of the Royal Society, London*.

TOLSTOY, J. 1971. *The Pulse of a Planet*. New York: New American Library.

WEXLER, H. 1951. Spread of the Krakatoa Volcanic Dust Cloud as Related to the High-Level Circulation. *Bulletin American Meteorological Society*, Vol. 32, No. 2, pp. 48–51.

Meteorite Impact

BEALS, C. S. 1971. Crustal Thickness and the Forms of Impact Craters. *Journal of Geophysical Research*, Vol. 76, No. 23, pp. 5586–95.

CHARTERS, A. C. 1960. High-Speed Impact. *Scientific American*, Vol. 203, No. 4, pp. 128–40.

FRENCH, B. M. and N. M. SHORT. eds. 1968. *Shock Metamorphism of Natural Materials*. Baltimore: Mono Book Corporation.

GLASS, I. I. April, 1973. Aerospace in the Next Century. *Institute for Aerospace Studies, University of Toronto, UTIAS Review No. 37.* See also May, June 1973. *Canadian Aeronautics and Space Journal*, pp. 193–215, 261–78. Kücheman, D. et al, eds. 1974. *Progress in Aerospace Sciences*, Vol. 15, New York: Pergamon Press (in press).

HEIDE, F. 1964. *Meteorites*. Chicago: University of Chicago Press.

MEEN, V. B. 1952. Solving the Riddle of Chubb Crater. *National Geographic Magazine*, Vol. C1, No. 1, pp. 1–32.

MILTON, D. J. et al. 1972. Gosses Bluff Impact Structure, Australia. *Science*, Vol. 175, No. 4027, pp. 1199–1207.
WEAVER, K. F. 1973. Have we Solved the Mysteries of the Moon? *National Geographic Magazine*, Vol. 144, No. 3, pp. 309–25.

Chapter 3

Bull Whip

BERNSTEIN, B., D. A. HALL, and H. M. TRENT, 1958. On the Dynamics of a Bull Whip. *Journal of the Acoustical Society of America*, Vol. 30, No. 12, pp. 1112–15.

Gunpowder

GREGORY, C. E. 1966. *Modern Explosives for Engineers*. Queensland: University of Queensland Press.
LAPP, R. E. 1968. *The Weapons Culture*. New York: W. W. Norton and Co.
WARDELL, W. H. 1892. *Encyclopedia Britannica*, Vol. 8, p. 806.

Nuclear Weapons

BRODE, H. L. 1964. A Review of Nuclear Explosion Phenomena Pertinent to Protective Construction. *Rand Corporation Report R-425-PR*.
BRODE, H. L., I. I. GLASS and A. K. OPPENHEIM. 1971. *Gasdynamics of Explosions Today*, in *Shock Tube Research*, Stollery, J. L. et al, eds. London: Chapman and Hall.
FOSTER, J. S., JR. 1970. Nuclear Weapons. *Encyclopedia Americana*. New York: Americana Corporation. Vol. 20, pp. 518–28.
GLASSTONE, S., ed. 1962. *Effects of Nuclear Weapons*. Washington: U.S. Atomic Energy Commission, Superintendent of Documents, U.S. Government Printing Office.

Sonic Boom

COLLINS, D. J. 1971. On the Experimental Determination of the Near-Field Behaviour of the Sonic Boom and its Application to Problems of N-Wave Focusing. New York: *AIAA 9th Aerospace Sciences Meeting, Paper No. 71-185*.
HUBBARD, H. N., chairman. Nov. 3, 1965. *Proceedings, Sonic Boom Symposium*. The Acoustical Society of America, St. Louis, Mo.
KANE, E. J. Nov. 3, 1965. Some Effects of the Nonuniform Atmosphere on the Propagation of Sonic Booms. *Proceedings, Sonic Boom Symposium, Acoustical Society of America*, St. Louis, Mo.
ONYEONWU, R. O. 1972. On the Effects of Wind and Temperature Gradients on Sonic-Boom Corridors. *Institute for Aerospace Studies, UTIAS Technical Note No. 168*. See also; July, 1973. Numerical Study of Aircraft Manoeuvers on the Focussing of Sonic Boom, *UTIAS Report No. 192*.
RIBNER, H. S., chairman and ed. 1972. in Sonic-Boom Symposium, Overview and Complementary Remarks. *Journal of the Acoustical Society of America*, Vol. 51, No. 2 (part 3), p. 672.
SYMONS, G. J., ed. 1888. The Eruption of Krakatoa and Subsequent Phenomena, *Report of the Krakatoa Committee of the Royal Society, London*.

Planetary Entry

BECKER, J. V. 1961. Reentry from Space. *Scientific American*, Vol. 204, No. 1, pp. 49–57.

D'ABELIO, G. F. and J. A. PARKER, 1971. *Ablative Plastics*. New York: M. Dekker Inc.

GLASS, I. I. 1961. Aerodynamics of Blasts. *Canadian Aeronautical Journal*, Vol. 7, No. 3, pp. 109–35.

HICKS,R. M., J. P. MENDOZA and F. A. GARCIA, JR., 1972. A Wind Tunnel-Flight Correlation of Apollo 15 Sonic Boom. *NASA TMX-62*, 111.

NAGLER, R. G. July, 1969. Tailoring Polymers for Entry into the Atmospheres of Mars and Venus. *Journal of Macromolecular Science-Chemistry*, A3(4), pp. 763–801.

SCHMIDT, D. L. May, 1969. Ablative Polymers in Aerospace Technology. *Journal of Macromolecular Science-Chemistry*, A3(3), pp. 327–65.

SHELDON II, C. S. 1971. A Brief Introduction to the Soviet Space Program, *AIAA Student Journal*, Vol. 9, No. 4, pp. 14–37.

VOJVODICH, N. S. May, 1969. Hypervelocity Heat Protection – A Review of Laboratory Experiments. *Journal of Macromolecular Science-Chemistry*, A3(3), pp. 367–94.

Chapter 4

Solar Wind

AZIMOV, I. 1968. *The Universe, from Flat Earth to Quasar*. New York: Discus Books (Avon).

BRANDT, J. C. 1970. *Introduction to the Solar Wind*. San Francisco: W. H. Freeman and Co.

BRICE, N. M. 1971. Our Outermost Atmosphere. Engineering, *Cornell Quarterly*, Vol. 5, No. 4, pp. 2–10.

HUNDHAUSEN, A. J. 1972. *Coronal Expansion and Solar Wind*. New York: Springer-Verlag.

LEVY, R. H., H. E. PETSCHEK and G. L. SISCOE. 1964. Aerodynamic Aspects of Magnetosphere Flow, *AIAA Journal*, Vol. 2, No. 12, pp. 2065–76.

ROEDERER, J. C. Jan., 1969. The Particles and Field Environment of the Earth, *Astronautics and Aeronautics*, Vol. 7, No. 1, pp. 22–28.

SPREITER, J. R. and A. Y. ALKSNE. 1970. Solar-Wind Flow Past Objects in the Solar System, *Annual Review of Fluid Mechanics*, Vol. 2, pp. 313–54. Palo Alto: Annual Reviews Inc.

SPREITER, J. R. Oct. 1972. Shock Waves in the Solar System. *Astronautica Acta*, Vol. 17, pp. 321–38.

Solar Flares

AZIMOV, I., 1968. *The Universe, from Flat Earth to Quasar*. New York: Discus Books (Avon).

COSWICK, R. and P. B. PRICE. 1971. Origins of Cosmic Rays, *Physics Today*, Vol. 24, No. 10, pp. 30–38.

FRIEDMAN, H. 1969. Solar Flares. *Astronautics and Aeronautics*, Vol. 7, No. 1, pp. 14–21.

GOLD, T. Aug., 1962. Cosmic Rays and the Interplanetary Medium. *Astronautics*, Vol. 7, No. 8, pp. 43–45.

HUNDHAUSEN, A. J. 1972. *Coronal Expansion and Solar Wind*. New York: Springer-Verlag.

Exploding Stars and Galaxies

AZIMOV, I. 1968. *The Universe, From Flat Earth to Quasar*. New York: Discus Books (Avon).

BOYER, S., E. T. BYRAM, T. A. CHUBB and H. FRIEDMAN. Aug., 1964. X-Ray Emission from the Crab Nebula. *Naval Research Reviews*, pp. 14–19.

BRANCAZIO, P. J. and A. G. W. CAMERON. 1969. *Supernovae and their Remnants*. New York: Gordon and Breach.

FRIEDMAN, H. Jan., 1972. Recent Progress and Future Prospects in High-Energy Astronomy. *Astronautics and Aeronautics*, Vol. 10, No. 1, pp. 24–28.

MIDDLEHURST, B. M. and L. H. ALLER. 1968. *Nebulae and Interstellar Matter*. University of Chicago Press, Chicago.

MILLMAN, P. M. 1965. *This Universe of Space*. Toronto: Canadian Broadcasting Corporation.

PENROSE, R. 1972. Black Holes. *Scientific American*, Vol. 226, No. 5, pp. 38–46.

REES, R. J. Oct., 1972. Extragalactic Explosive Phenomena. *Astronautica Acta*, pp. 315–320.

SANDAGE, A. R. 1964. Exploding Galaxies, *Scientific American*, Vol. 211, No. 5, pp. 38–47.

SEDOV, L. I. 1959. *Similarity and Dimensional Methods in Mechanics*. New York: Academic Press Inc.

SPREITER, J. R. Oct., 1972. Shock Waves in the Solar System. *Astronautica Acta*, Vol. 17, pp. 321–38.

ZWICKY, F. Oct., 1972. Morophology of Rapid Cosmic Processes. *Astronautica Acta*, pp. 307–313.

Cosmological Big Bang

ARP, H. Dec., 1971. Observational Paradoxes in Extragalactic Astronomy. *Science*, Vol. 174, No. 4015, pp. 1189–1200.

AZIMOV, I. 1968. *The Universe, from Flat Earth to Quasar*. New York: Discus Books (Avon).

Chapter 5

GREGORY, C. E. 1966. *Modern Explosives for Engineers*. Queensland: University of Queensland Press.

JACOBSEN, S. May, 1972. Turning up the Gas. *Bulletin of the Atomic Scientists*, pp. 35–38.

LOUGH, T. S. June, 1968. *Peaceful Nuclear Explosions and Disarmament*. Peace Research Review.

MOSES, S. A. 1966. Electroexplosive Devices in Aerospace Vehicle Systems. *IEEE 1966 Aerospace Conference*, Seattle, Washington.

NORDYKE, M. D. 1970. Technical Status Summary of Peaceful Uses for Nuclear Explosives. *Lawrence Radiation Laboratory, UCRL-72332*.

PATTERSON, A. M. and J. M. DEWEY, 1970. Review of Fireball/Shock Wave

Anomalies on TNT Charges Detonated at Suffield from 1958 to 1969. *DRES Suffield Technical Note No. 275.*

REYNOLDS, G. June–July, 1958. Ripple Rock – The End Comes with a Bang. *du Pont Magazine.*

RUBIN, B., L. SCHWARTZ and D. MONTAN, 1972. An Analysis of Gas Stimulation Using Nuclear Explosives. *Lawrence Livermore Laboratory, UCRL-51226.*

TELLER, E. et al. 1968. *The Constructive Uses of Nuclear Explosives.* New York: McGraw-Hill Book Co.

TOMAN, J. 1970. Summary of Nuclear-Excavation Applications. *Lawrence Radiation Laboratory Preprint UCRL-72220.*

Chapter 6

BRODE, H. L. 1964. A Review of Nuclear Explosion Phenomena Pertinent to Protective Construction. *Rand Corporation Report R-425-PR.*

——— 1959. Blast Wave from a Spherical Charge. *Physics of Fluids*, Vol. 2, No. 2, p. 217.

COLE, ROBERT H. 1948. *Underwater Explosions.* Princeton: Princeton University Press.

GLASS, I. I. 1961. Aerodynamics of Blasts. *Canadian Aeronautical Journal*, Vol. 7, No. 3, pp. 109–35.

——— and G. N. PATTERSON, 1955. A Theoretical and Experimental Study of Shock-Tube Flows, *Journal of the Aeronautical Sciences*, Vol. 22, No. 2, pp. 73–100.

KINNEY, G. F. 1962. *Explosive Shocks in Air.* New York: Macmillan Co.

Chapter 7

ALCOCK, A. J. Nov.–Dec., 1970. Laser-Produced Plasma at NRC's Physics Division *Canadian Research and Development*, Vol. 3, No. 6, pp. 19–22.

BRODE, H. L., I. I. GLASS and A. K. OPPENHEIM. 1971. *Gasdynamics of Explosions Today*, in *Shock Tube Research*; Stollery, J. L. et al, eds. London: Chapman and Hall.

GLASS, I. I. 1967. Research Frontiers at Hypervelocities. *Canadian Aeronautics and Space Institute Journal*, Vol. 13, Nos. 8 and 9, pp. 347–66 and 401–26.

KINSLOW, R. 1970. *High Velocity Impact Phenomena.* New York: Academic Press Inc.

Chapter 8

BOYER, A. G. May, 1964. Design, Instrumentation and Performance of the UTIAS 4 in. × 7 in. Hypersonic Shock Tube. *Institute for Aerospace Studies, University of Toronto*, UTIAS Report. No. 99.

CHAN, S. K. August, 1973. An Analytical and Experimental Study of an Implosion-Driver Shock-Tube. *Institute for Aerospace Studies, University of Toronto*, UTIAS, Report No. 191.

CHAN, Y. Y., R. P. MASON and N. M. REDDY, June, 1965. Instrumentation and Calibration of UTIAS 11 in. × 15 in. Hypersonic Shock Tunnel. *Institute for Aerospace Studies*, University of Toronto, Technical Note No. 91.

ENKENHUS, K. R. June, 1957. The Design, Instrumentation and Operation of the UTIA Low Density Wind Tunnel. *Institute for Aerospace Studies, University of Toronto, UTIAS* Report No. 44.

FLAGG, R. F. June, 1967. The Application of Implosion Wave Dynamics to a Hypervelocity Launcher. *Institute for Aerospace Studies, University of Toronto*, UTIAS Report No. 125.

FRENCH, J. B. and E. P. MUNTZ, March, 1960. Design Study of the UTIA Plasma Tunnel. *Institute for Aerospace Studies, University of Toronto*, Technical Note No. 34.

GLASS, I. I. 1972. *Appraisal of UTIAS Implosion-Driven Launchers and Shock Tubes, in Progress in Aerospace Sciences*; Küchemann, D., et al, eds. Vol. 13 Oxford: Pergamon Press.

——— 1967. Research Frontiers at Hypervelocities, *Canadian Aeronautics and Space Institute Journal*, Vol. 13, Nos. 8 and 9, pp. 347–66 and 401–26.

———, H. L. BRODE, and S. K. CHAN, March, 1974. Strong Planar Shock Waves Generated by Explosively-Driven Spherical Implosions. *AIAA Journal*, Vol. 12, No. 3, pp 367–74.

——— 1961. Aerodynamics of Blasts. *Canadian Aeronautical Journal*, Vol. 7, No. 3, pp. 109–35.

——— and L. E. HEUCKROTH, 1963. Hydrodynamic Shock Tube. *Physics of Fluids*, Vol. 6, No. 4, pp. 543–47. 1968. Low-Energy Underwarer Explosions, *Physics of Fluids*, Vol. 11, No. 10, pp. 2095–107.

——— and G. N. PATTERSON. A Theoretical and Experimental Study of Shock-Tube Flows, *Journal of the Aeronautical Sciences*, Vol. 22, No. 2, pp. 73–100.

———, H. S. RIBNER, and J. J. GOTTLIEB, Oct. 1972. Canadian Sonic-Boom Simulation Facilities. *Canadian Aeronautics and Space Journal*, pp. 235–46. See also 1972 ICAS paper No. 72–26.

LAFRANCE, J. C. July, 1966. A Study of a Neon Cycle Cryogenic Pumping System for a Low Density Plasma Tunnel. *Institute for Aerospace Studies., University of Toronto.* Report No. 116.

RANGI, R. S. August, 1964. Information for Users of the National Research Council's 5-Ft × 5-Ft Trisonic Blowdown Wind Tunnel at the National Aeronautical Establishment, Uplands. *N.A.E.* Misc. 34.

Chapter 9

BERNAL, J. D. 1969. *The World, The Flesh and the Devil*. Bloomington: Indiana University Press.

FORRESTER, J. W. 1971. Churches at the Transition between Growth and World Equilibrium. *Massachusetts Institute of Technology*.

GABOR, D. 1972. *The Mature Society*, New York: Praegar, Publishers, Inc.

GLASS, I. I. April, 1973. *Aerospace in the Next Century*. Institute for Aerospace Studies, University of Toronto. UTIAS *Review No. 37*, See also, May, June, 1973. *Canadian Aeronautics and Space Journal*, pp. 193–215, 261–78; Küchemann, D. et al, eds. 1974. *Advances in Aerospace Sciences*, Vol. 15, New York: Pergamon Press (in press).

Soncino Books of the Bible. 1968. London: The Soncino Press.

THOMPSON, W. I. 1971. *At the Edge of History*. New York: Harper & Row.

WOOLDRIDGE, D. E. 1969. *Mechanical Man – The Physical Basis of Intelligent Life*. New York: McGraw-Hill.

YEHOASH (Bloomgarden). 1957. *Yiddish Translation of the Bible*. New York: "Forward" edition. Yehoash Farlag Gezelshaft.

SOME RECOMMENDED TECHNICAL REFERENCE BOOKS

CHERNYI, G. G. 1961. *Introduction to Hypersonic Flow*. New York: Academic Press.

COURANT, R. and K. O. FRIEDRICKS. 1948. *Supersonic Flow and Shock Waves*. New York: Interscience Publishers, Inc.

GAYDON, A. G. and I. R. HURLE. 1963. *The Shock Tube in High-Temperature Chemical Physics*. London: Chapman and Hall Ltd.

GLASS, I. I. and J. G. HALL. 1959. *Shock Tubes, Handbook of Supersonic Aerodynamics, Section 18, Navord Report 1488 (Vol. 6)*, Washington: U.S. Government Printing Office.

KINSLOW, R. ed. 1970. *High-Velocity Impact Phenomena*. New York: Academic Press.

LIEPMANN, H. W. and A. ROSHKO, 1957. *Elements of Gasdynamics*. New York: John Wiley and Sons, Inc.

LUKASIEWICZ, J. 1973. *Experimental Methods of Hypersonics*. New York: Marcel Dekker, Inc.

MARTIN, J. J. 1966. *Atmospheric Reentry*. Englewood Cliffs, N.J.: Prentice Hall Inc.

POPE, A. and K. L. GOIN. 1965. *High Speed Wind Tunnel Testing*. New York: John Wiley and Sons, Inc.

SHAPIRO, A. H. 1953. *The Dynamics and Thermodynamics of Compressible Fluid Flow*. New York: Ronald Press Co.

ZEL'DOVICH, YA. B. and YU. P. RAIZER. 1966. *Physics of Shock Waves and High Temperature Hydrodynamic Phenomena*. New York: Academic Press.